Hamza Yassin

Hamza's Wild World

Illustrated by
Louise Forshaw

MACMILLAN CHILDREN'S BOOKS

Published 2024 by Macmillan Children's Books
an imprint of Pan Macmillan
The Smithson, 6 Briset Street, London EC1M 5NR
EU representative: Macmillan Publishers Ireland Ltd, 1st Floor,
The Liffey Trust Centre, 117–126 Sheriff Street Upper
Dublin 1, D01 YC43
Associated companies throughout the world
www.panmacmillan.com

ISBN 978-1-0350-3221-1

1 3 5 7 9 8 6 4 2

A CIP catalogue record for this book is available from the British Library.

Printed and bound by CPI Group (UK) Ltd, Croydon CR0 4YY
Design by Janene Spencer

Photos from Hamza's personal collection on pages: 2, 3, 4, 7, 8, 10, 11, 17, 41,
65, 86, 103, 123, 124, 125, 129, 138, 138, 139, 139, 161, 162, 163, 175, 184, 189, 193,
195, 203, 215, 224, 225, 226, 227, 228, 229, 255, 285, 286, 286, 307, 316, 335, 342;
Page 118 © Dr Gemma Clucas; **Page 186** © Erin Ranney;
Page 217 © Arterra Picture Library/Alamy; **Page 241** © Everett Collection Inc/Alamy;
Page 275 third photo down on right-hand-side © Catherine Brereton;
Page 277 © Liam WhiteAlamy; **Page 391** © John Sparks/Nature Picture Library;
Page 393 © Wikimedia; All other images © Shutterstock

Contents

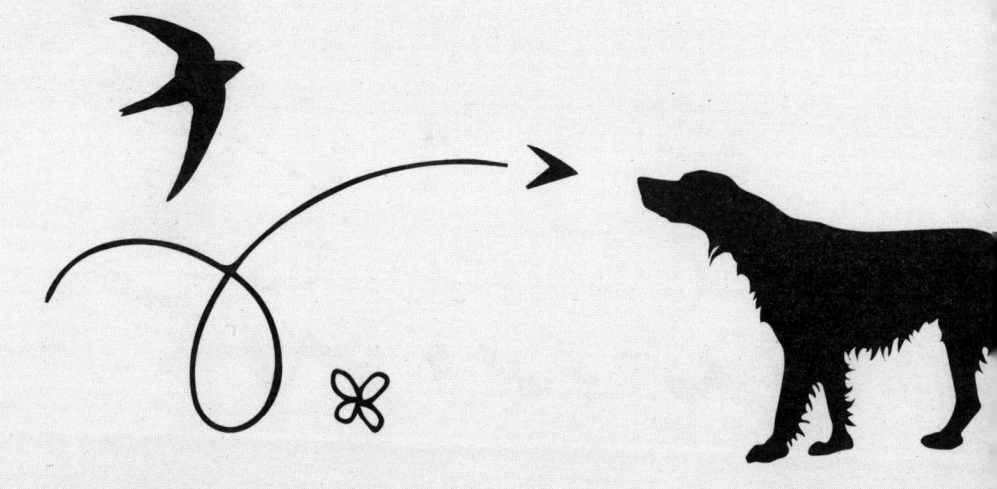

I have talked about my dyslexia in this book so you may wonder how I managed to write a book. There are many ways to write a book and anyone can be an author.

This book was written over many months. I had weekly meetings with my writer – Catherine Brereton – and we talked about all of the animals we wanted to include in the book, the best facts and my experiences of filming and interacting with the animals and landscapes.

We transferred all the recordings to the page and then spent our weekly meetings shaping the text, adding more facts and we consulted with Dyslexia Scotland to help make this as accessible as possible. I truly believe dyslexia is my superpower. If you have a story to tell you will find a way to tell it.

A Little Bit About Me

My name is Hamza Yassin. You might know me as a
wildlife presenter, or maybe from *Strictly Come Dancing*.
My main job, though, is being a cameraman. I spend most
of my time watching and filming amazing wildlife – in the
west of Scotland, where I live, and in incredible places
around the world.

I was born in Sudan and I moved
to England aged eight. I didn't
know a word of English at that
point. It was December, so
one of the first things I saw
everywhere was this big guy
dressed in a red jacket
saying 'Ho ho ho!'
Which I didn't
really understand!

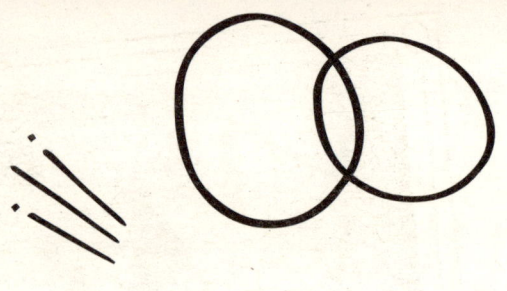

The rest of my family could speak some English, but not me. My parents thought I needed to start learning English as soon as possible, so they turned on the TV for me to get used to hearing the language. The first thing they turned on was Sir David Attenborough's *Life of Birds*. Wow. I loved it. I recommend you take a look. And the next thing I saw was Steve Irwin, shouting, 'G'day, mate,' and wrestling a crocodile. This blew my mind. In Sudan we were taught never to go near a crocodile – there are loads of myths about how they'll grab you and bury you and eat you – but here someone was wrestling it? Amazing. I was hooked, and David Attenborough and Steve Irwin became two

 of my heroes. I wanted to be like them. And I thought, if I couldn't be like them, I could be the cameraman who films them. This is how my passion and ambition started.

I got my first camera aged twelve, and I knew then that I loved photography. I'm a happy snapper: I'll take photos of anything and everything. One of the first wildlife things I tackled was the ducks at the local park. It was easy, with everyone feeding them. There were mallards, geese, swans, cygnets, ducklings. Great.

Around the same time, I asked a whole load of cameramen and women, 'How do I become a wildlife cameraman?' Everybody had a different route to eventually becoming a camera person.

I never thought I would do it, though, because my whole family are doctors and dentists and I thought they would

3

expect me to do this, too. I put pressure on myself, to be perfectly honest. I would tell my family I wanted to be a dentist like my brother.

I do photography because I love it, but also filming is visual, which is perfect for me because I don't read and write very well. Even as a little boy in Sudan I struggled at school. I knew that there was something wrong. I hoped that maybe I just had a problem with Arabic, but it was the same with English. I didn't do well for a long time in school.

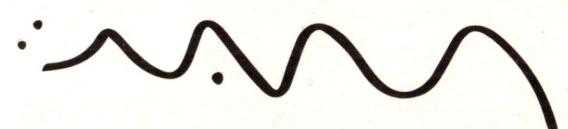

It was a teacher, Mrs Strange, who suggested I might have dyslexia. When I first heard the word I was really worried. But being dyslexic simply means finding it very hard to read and write. It didn't mean I wasn't bright – just that I needed special help with reading and writing. So my school got me a reader and a scribe (someone who writes my thoughts down for me in exams), and I shot from near the bottom to near the top of the class.

Now I think of dyslexia as my superpower. I'm an ambassador for Dyslexia Scotland and I'm really proud and honoured to be so. It's great to show people how amazing it can be to have dyslexia. Einstein was dyslexic. I see patterns really well – for example, I don't need directions when I'm driving and I know my birds very well.

I got As and Bs in my GCSEs and A-levels and I was all set to go to university to study dentistry, following in my family's footsteps. But I was unhappy because I really didn't want to do it. I wanted to be a cameraman. I plucked up the courage to tell my parents this and they were completely supportive. 'Follow your dreams,' they said.

I started emailing those cameramen and women I had spoken to and they said, for the BBC to take me seriously as a wildlife cameraman I would need a zoology degree. So this is what I did – zoology with conservation and animal behaviour – and I loved it. That's where I met the great wildlife cameraman Jesse Wilkinson. After uni I messaged Jesse online for weeks and weeks. One day I saw him at an RSPB reserve and I persuaded him to put my number in his phone. I said, 'I'm a big rugby player, I can work for you, I can carry your kit for you.' Jesse wasn't keen – he said he normally works alone . . . and the

teams are very small . . . it's hard to just bring a random
person along . . .

But a few weeks later I got a call. It was Jesse, asking if he
could take me up on my offer of carrying his kit, because
he'd hurt his back, and he was on a shoot, and the shoot
was around Scotland! There was no budget to pay me but
I'd be saving them from having to cancel the shoot, and
my food and accommodation would be paid for. Great,
I thought, I get to learn from the best.

That shoot took three weeks, and in that time Jesse realized I really knew my birds. I just needed a break to get into the industry. I had found a crack in the door, and wedged my foot right in! I then became Jesse's assistant for a number of years, doing my own bits of filming when Jesse was on his breaks, and eventually I became a fully fledged cameraman in my own right. I moved to Scotland and the rest is history.

If you have a dream, you can reach it. I knew I wanted to be a wildlife cameraman. Presenting wasn't so much on my radar but I knew I couldn't be a researcher, writer or producer. I can't read or write very well but I can get 3 metres away from an otter and the otter doesn't know I'm there! That's my special skill.

I moved to the west coast of Scotland because that's where a lot of the shoots happen. I cleaned houses, chopped logs, cut grass, moved furniture, became a general odd-job man. I ate at the same pub every night, and showered at the community centre. Sometimes I jumped in the sea to save myself the one pound it cost to have a shower, until I had enough work and enough money to get a proper place to live.

Working on *Wild Isles* with Jesse and John Aitchison was a dream come true for me. Now I always say, follow your dream. My dance partner Jowita would say the same! If you want to be a camera person, you can be. You can start off with just your phone. Want to go underwater? You can use a GoPro – they're cheap. Start taking your own photos, find out what you like to do best and use your skills. Whatever equipment you have, it's all about understanding the animals and where and when to find them.

Human Stats

We humans are part of the animal world and share many things with our fellow animals. We get hungry and frightened. We sense the world around us. We run, jump, walk, swim and sleep. We grow from babies to adults. On the other hand, it is fascinating to see how different animals can be. We cannot fly like birds, we are not as hairy as bears, and we don't have feathers, scales, wings or horns. We don't lay eggs. Some animals have more eyes than us, can hear much better than us or have more hearts than us. The animal world is full of variety.

Did you know?

Modern humans appeared on the Earth around 300,000 years ago in east Africa. We are all descendants of these common ancestors.

Facts

Scientific name:	*Homo sapiens*
Mammal family:	great apes (along with bonobos, chimpanzees, gorillas and orangutans)
Height:	world average 171 cm for a man, 160 cm for a woman, 149 cm for a ten-year-old child, but individuals can be much taller or shorter than this
Found:	all over the world – 7.89 billion of us!
Eats:	all sorts of meat and plant food
Babies:	an individual woman has between 0 and over 40 babies. The world average is 2.31 per woman
My three words:	love, empathy, forward-thinking

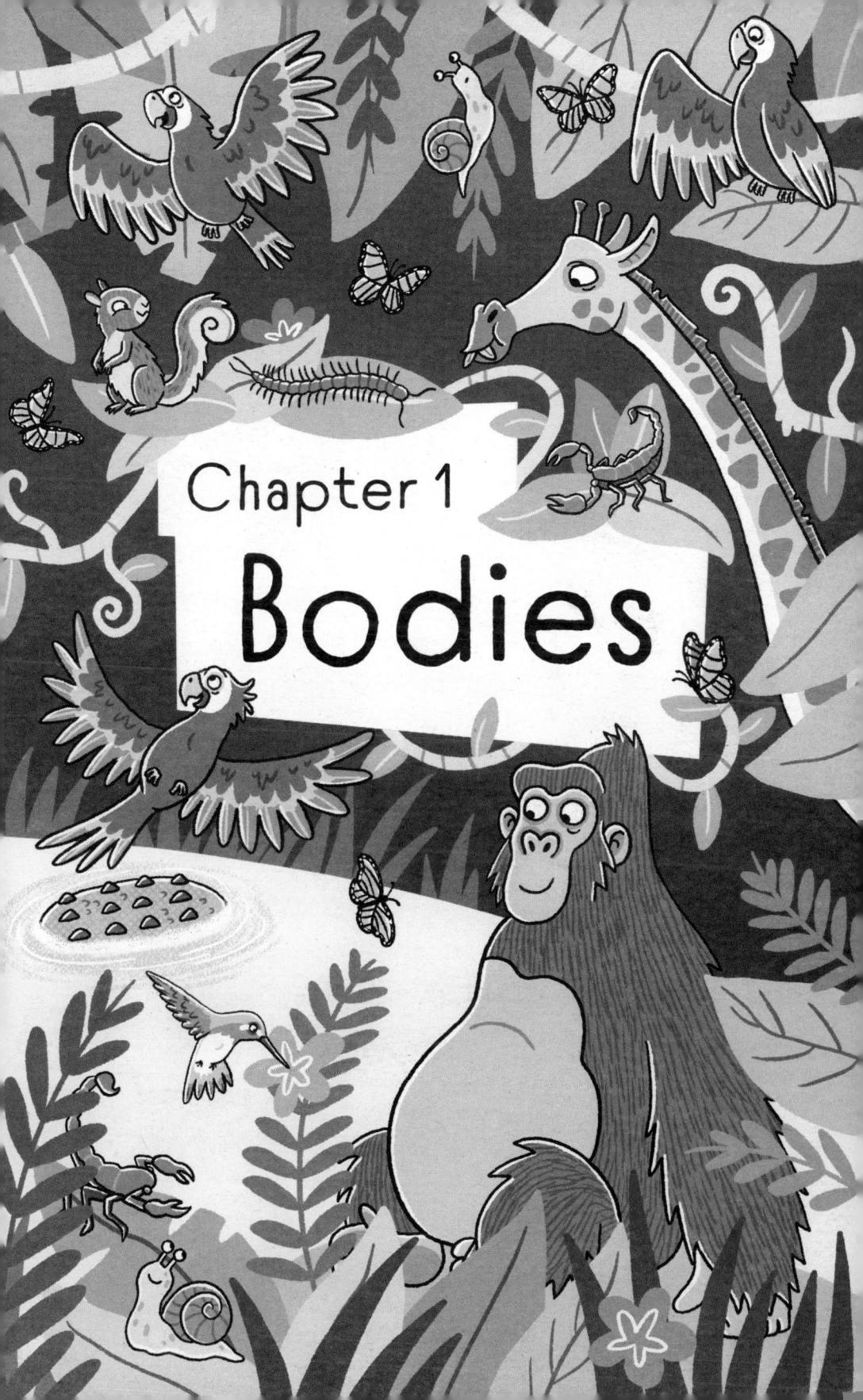

Chapter 1
Bodies

Now it's time for a countdown of my top ten favourite animals. In at number ten, it's the **sparrowhawk**. It's a beautiful bird of prey. It's one of the most common birds of prey in the UK and one that most people will be able to see.

The sparrowhawk is special to me because it's one of the first things that got me noticed at the BBC. It was when I started working for *The Hunt*, a programme which took a detailed look at predators and their prey – it was this bird that gave me my first big break into the industry.

I said to Jesse Wilkinson, 'I've got this nest.' It was a sparrowhawk nest that I'd found. The clue that gave it away was a little tree stump that had been chopped off. It looked like a butcher's block, with legs and guts and feathers and so

on hanging off it. I started taking pictures of it. It was the middle of summer and I get severe hay fever but I basically stuffed tissues up my nose and sat in the hide for three weeks while I took over 2,500 images. Out of all of them there's only one image that I really love. It's of a male sparrowhawk with a twinkle in his eye – but he's facing the wrong way. I've gone back year after year to try and get that image again but even better – I've still not got it.

Even so, I'd got all these lovely shots. I took them to Jesse and said, 'Look. I've got this great shot of a sparrowhawk.'

He wanted to know where it was. I didn't want to tell him –
it was my discovery, my filming spot. But he said he was
doing a film for Sir David Attenborough. Wow! So, OK,
I told Jesse the location so he could film it. But I asked him
to tell his bosses that it was me who found it, me who got
the first shots.

I have lots of sparrowhawk stuff at my house, like a skull
and a stuffed bird. For me, it's just incredible how clever
they are. For example, they watch you fill up the bird
feeders and they realize what's going on, then they fly
really low to the ground behind the fence and out of sight,
then swoop over the top of it like it's a McDonald's Drive
Thru . . . and just pick up their food order. The small bird
on the feeder is caught almost before it knows what's
happening, caught in the sparrowhawk's strong talons
and swept off and away in one swift movement.

Sometimes it launches itself from a perch and takes just
a few seconds to swoop and hit the prey. In the swoop it

Whoosh! I can fly very fast . . . small birds watch out!

reaches a flying speed of up to
50 kilometres an hour. It can also turn in
mid-air, avoiding crashing into a tree or wall.

Once they've caught it, sparrowhawks take their prey to a
perching spot or a plucking place to kill and eat it.

They pluck the breast feathers out and then tear the meat
with their beak. You can tell a pile of feathers is the result
of a sparrowhawk kill rather than a fox or a cat because the
shafts of the larger feathers will be plucked clean out of the
skin, with the perfect round end intact and not broken.

You can time when they're going to arrive in your
garden – they're looking for when all the small birds are
chattering ready to go to roost.

Blue tits have a particular alarm call that means
'sparrowhawk nearby' and several other types of birds
can understand this call.

Find out more: Skulls, page 224; feathers, page 137

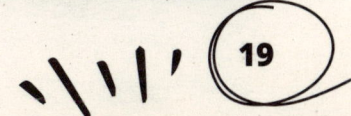

Facts

Scientific name:	*Accipiter nisus*
Bird family:	hawks, eagles, kites and vultures
Height:	28–38 cm
Wingspan:	55–70 cm
Found:	woodland, gardens
Eats:	mainly small birds such as finches, tits and sparrows, rodents, pigeons
Eggs:	4–5 per clutch
My three words:	fierce, beautiful, hunter

The sparrowhawk has white 'eyebrows' which make it look fierce.

Sparrowhawks are well camouflaged, with a dark back and patterns of bars on their front that confuse the eye. The male has a grey back and the female has a brown one. The male has a beautiful rosy colour on his breast.

As with most birds of prey, the females are quite a lot larger than males. The females need to be bigger and stronger to have enough body reserves for laying eggs and the males need to be smaller and more agile because they do most of the hunting for a nesting pair. Also, they both hunt different prey – the male goes for the smaller, tweety birds (such as finches, tits, sparrows) and the female goes for the slightly larger birds, such as pigeons or thrushes. So together they broaden their range of food.

Animal bodies

Animal bodies are built in just the right way for the job –
for survival and success.

Gut

An animal's gut is a long tube from its mouth to the other
end. Food goes in one end and waste comes out the other.

Skeleton

Some skeletons are made of bone. Some are made of
cartilage – that's the gristly stuff you can feel at the
end of your nose. Some are a hard shell and
some are made of water. Skeletons keep
an animal's body in shape. Some are
inside the animal's body, some are on
the outside.

Muscles

Muscles keep the animals on the move. An ant's muscles
may be tiny but they are surprisingly strong. The blue
whale has one of the biggest (and strongest): the peduncle,

which stretches from the whale's back fin to its fluke (or tail). This can be 7.5 metres long.

Heart and blood

Most animals have hearts pumping blood around their bodies to keep them alive. Sea stars (sometimes called starfish) have no heart or blood at all. They pump seawater around inside their bodies instead.

Breathing system

Reptiles, amphibians, mammals and birds breathe using their lungs. Fish and crabs use gills. Insects and spiders breathe using little tubes all over their bodies. Amphibians breathe through their skin too. Whatever works best for them!

Brain

The brain is the control centre in charge of movement, the senses, thoughts, memories and emotions. It also deals with breathing, digestion, sleeping and keeping warm. That's a lot for one body part to do. I think of the brain as like the conductor of an orchestra. It needs every other body part's help in order to work. All the sections are needed for the body to run smoothly.

How many legs?

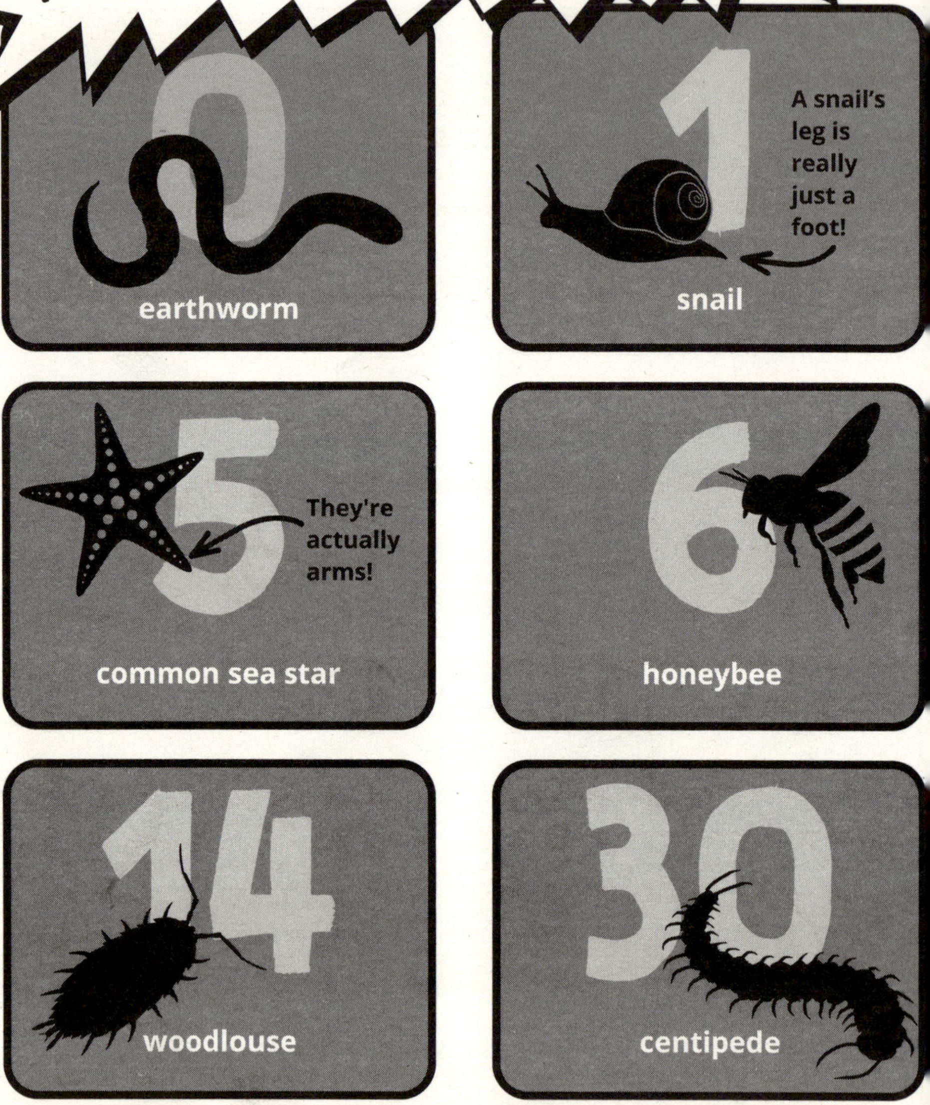

0

earthworm

1

A snail's leg is really just a foot!

snail

5

They're actually arms!

common sea star

6

honeybee

14

woodlouse

30

centipede

pigeon

zebra

scorpion

crab

millipede

A slow worm is not actually a worm. It is a lizard that looks like a snake. But sometimes, if you look closely, it has tiny little stumps for legs.

Animals with backbones

Let's look at some different types of animal bodies. One big category – which happens to include all of my top ten favourite animals – is the **vertebrates**. These are animals with backbones and skeletons made of bone or cartilage. The skeleton is inside the body and is called an **endoskeleton**.

Fish

Fish live in water. Some have skeletons made of cartilage and others have skeletons made of bone. They have fins instead of legs. They breathe through gills. Sharks, rays, eels, minnows, seahorses, salmon and clownfish are some of the many types of fish.

Amphibians

Most amphibians like to live in, or near, water. Adults have legs and breathe with lungs and through their skin. Toads, frogs, newts, salamanders and the axolotl are all amphibians.

Reptiles

Reptiles have dry, scaly skin and are cold-blooded, which means they need heat from the sun to warm up each day. Snakes, lizards, crocodiles and turtles are all reptiles.

Birds

I'm known for being a birdwatcher, and I just love birds! One of many amazing things about birds is that they have wings and feathers, although not all of them can fly. They lay eggs and are warm-blooded, which means they make their own warmth inside of their bodies.

Mammals

Mammals are warm-blooded, have fur or hair, and have four limbs. Most mammals don't lay eggs; they give birth to their young instead, and feed them with milk made in the mother's body. We are mammals – and so are monkeys, tigers, rabbits, wolves, whales, seals, zebras, bats, mice, kangaroos and many, many more.

1

Blue whale
30 m long
15 Hamzas

2

Whale shark
18 m long
10 Hamzas

3

Giant squid
14 m long
7 Hamzas

4

Saltwater crocodile
7 m long
3.5 Hamzas

5

Giraffe
5.7 m tall
2.85 Hamzas

6

Hippopotamus
5 m long
2.5 Hamzas

7

African elephant
4 m long
2 Hamzas

8

Ostrich
2.75 m tall
1.5 Hamzas

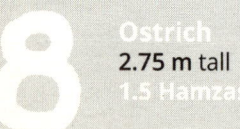

Ten big animals

9 Polar bear
2.5 m long
1.25 Hamzas

10 Grizzly
(brown) bear
2.1 m long
1 Hamza

The biggest animal

The blue whale is the biggest animal there has ever been on Earth – even bigger than any dinosaurs. It tops the table for lots of other **biggest** things as well . . .

- **Biggest** heart – its heart is enormous, measuring more than 1.5 metres long and weighing 180 kilograms – that's the same as 600 human hearts. Its biggest blood vessel is called the aorta and is about as wide as your head!

- **Biggest** tongue – which weighs as much as an elephant.

- **Biggest** muscle – the peduncle (see page 22).

- **Biggest** baby – a newborn blue whale is as big as an adult bull elephant. A baby drinks a bathful of milk a day.

- **Biggest** sound – blue whales are the loudest animals. They make deep groans and moans that can be heard by other whales up to 1,600 kilometres away.
- The female is bigger than the male.
- They eat really small things. Their main food is krill, tiny shrimp-like animals. They strain them out of the water using their baleen plates, which are like big bristly combs inside their mouths, and they eat BIG portions – about 6 tonnes a day.
- Blue whales are found in all oceans except the Arctic.
- Whales with baleen are called baleen whales or moustache whales. This is one of two kinds of whales. The others are toothed whales (including dolphins).

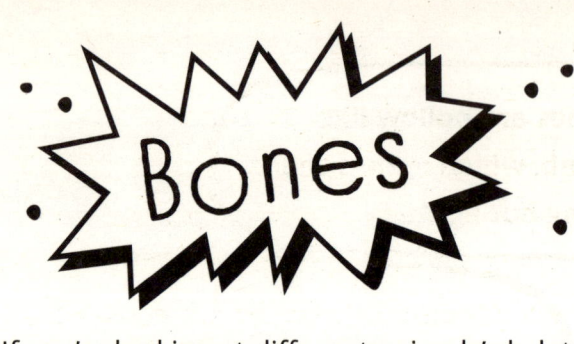

Bones

What do skeletons say before they start a meal?
Bone appétit!

If you're looking at different animals' skeletons you'll realize we're not all that different under the skin.

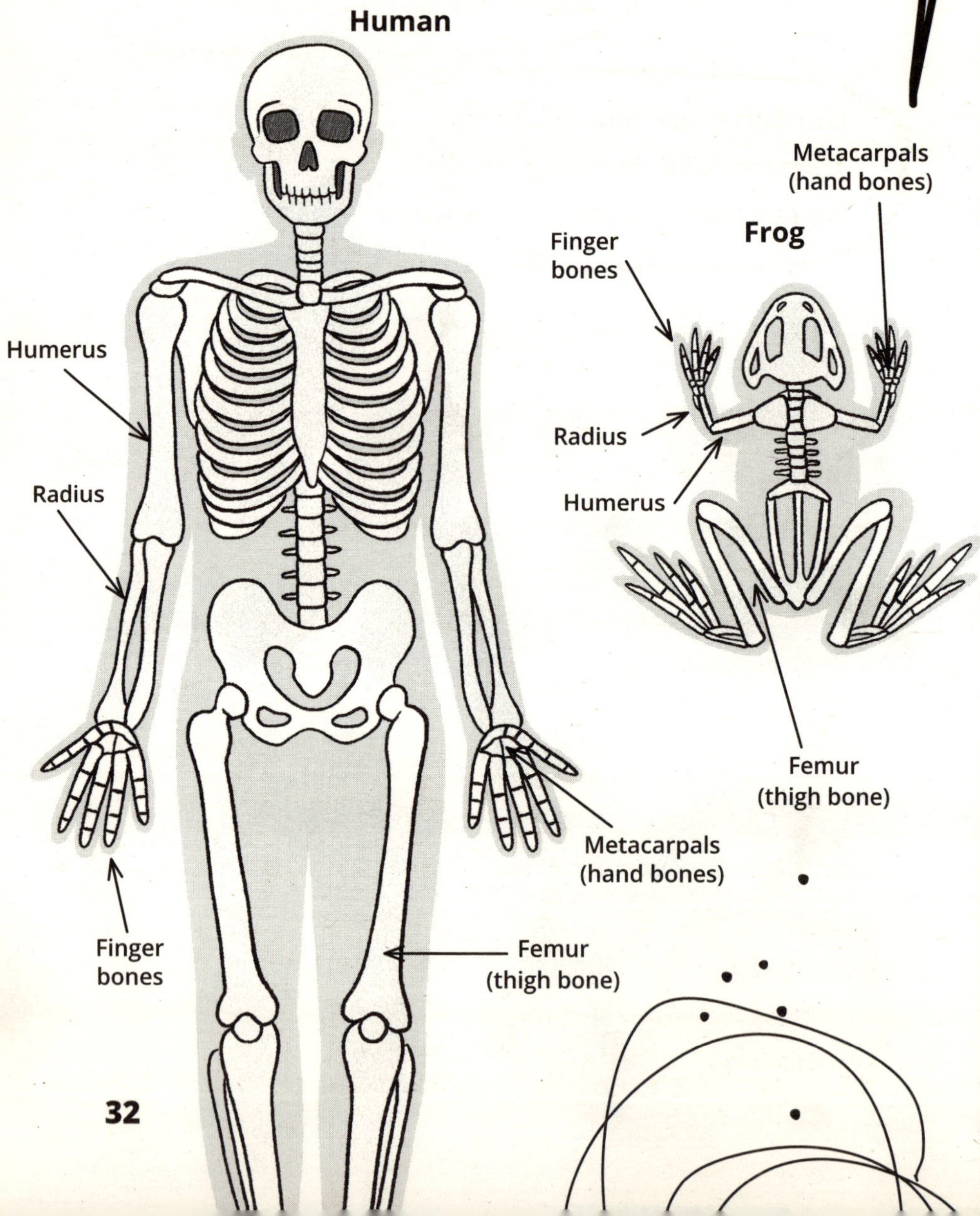

Human

Humerus

Radius

Finger bones

Metacarpals (hand bones)

Femur (thigh bone)

Frog

Finger bones

Metacarpals (hand bones)

Radius

Humerus

Femur (thigh bone)

Birds' bones are hollow like honeycomb, which makes them very strong but light.

Chimpanzee bones are denser than ours, but also spongier, so their bodies can cope with the stress of swinging and dropping from branches.

Our bones are not as strong as chimpanzees', but they are lighter and we can float (just about!) in water.

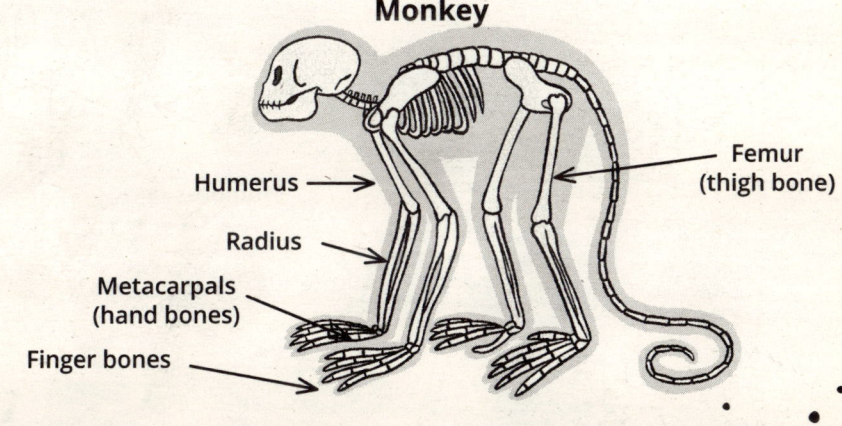

Monkey

Humerus

Radius

Metacarpals (hand bones)

Finger bones

Femur (thigh bone)

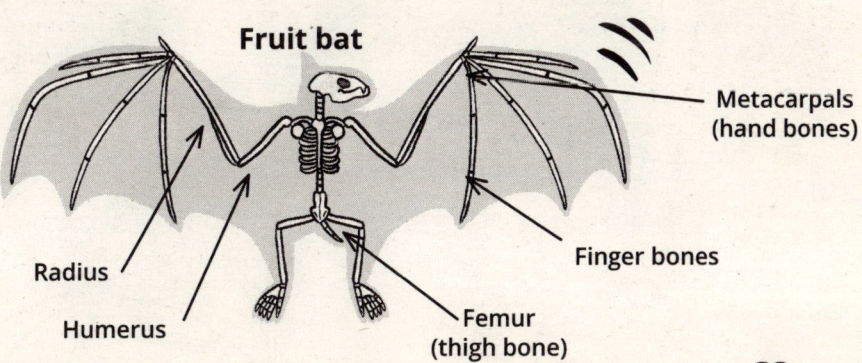

Fruit bat

Radius

Humerus

Femur (thigh bone)

Finger bones

Metacarpals (hand bones)

Muscles

Animals' muscles are different shapes and sizes for the different lives they lead.

Biceps

Pectoral

Triceps

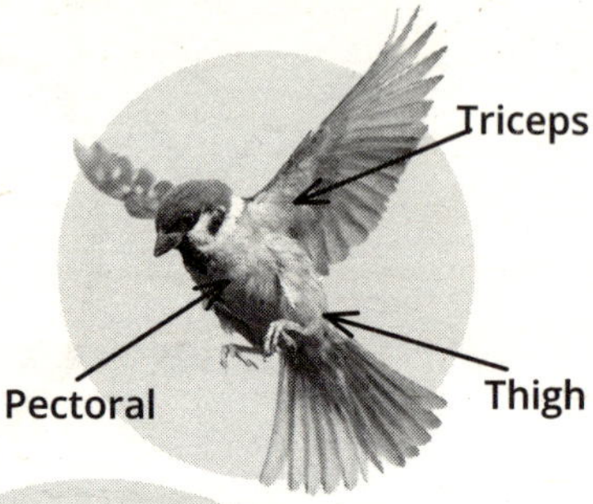

Triceps

Pectoral

Thigh

Pectoral

Triceps

Biceps

Quadriceps

Hamstring

34

An elephant has up to 40,000 muscles just in its trunk – that's more than you have in your whole body.

Birds have muscles called the supracoracoideus muscles, which pull their wings upwards. Humans and other mammals have two smaller shoulder muscles instead.

Strong muscles aren't just for animals with backbones. Ants' muscles allow them to lift things 20 times their body weight. That's like you lifting 20 of your classmates at once.

The silverback gorilla has a little crown that he gets as part of being the silverback. It's actually a muscle, that only develops as the gorilla gets older.

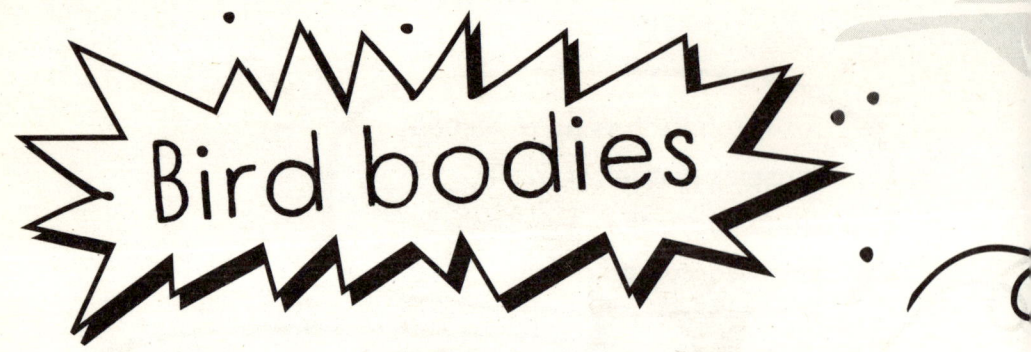

Bird bodies

Bird bodies are built for flight. Here are a few basic features:

- ◎ **Wings** – shaped to lift them up in flight.
- ◎ **Bones** – strong with a honeycomb structure. The main wing bone is not solid like in a mammal but light and strong, with air sacs. Scientists now think that birds can let more air into the spaces in their bones, making them even firmer and stronger – like an inflatable boat when it's fully blown up.
- ◎ **Feathers** – have several jobs but the main one is to make birds even more efficient at flying.
- ◎ Strong **chest muscles** to work their wings. The muscles are attached to the keel, which is a ridge of bone along the centre of the breastbone.
- ◎ A **streamlined body** for efficiency in flight.
- ◎ A **beak** instead of heavy jaws and heavy teeth.

Of course, there are a few birds that can't fly, but they all still have wings. Penguins use their wings to swim. Kiwis' wings are hidden away.

Bird colours

Birds are often intricately coloured. Even an all-brown bird might have beautiful, detailed patterns on its feathers. For a birdwatcher, learning to spot and describe these colours and patterns is a big part of identifying birds. Maybe we'll look out for a yellow rump, a red crown or a mottled back, for example. This is where it comes in useful to know the names for the parts of a bird's body.

Crown

Nape

Forehead

Mantle

Back

Tail

Rump

Chin

Throat

Primary feathers

Secondary feathers

Breast

Flank

Belly

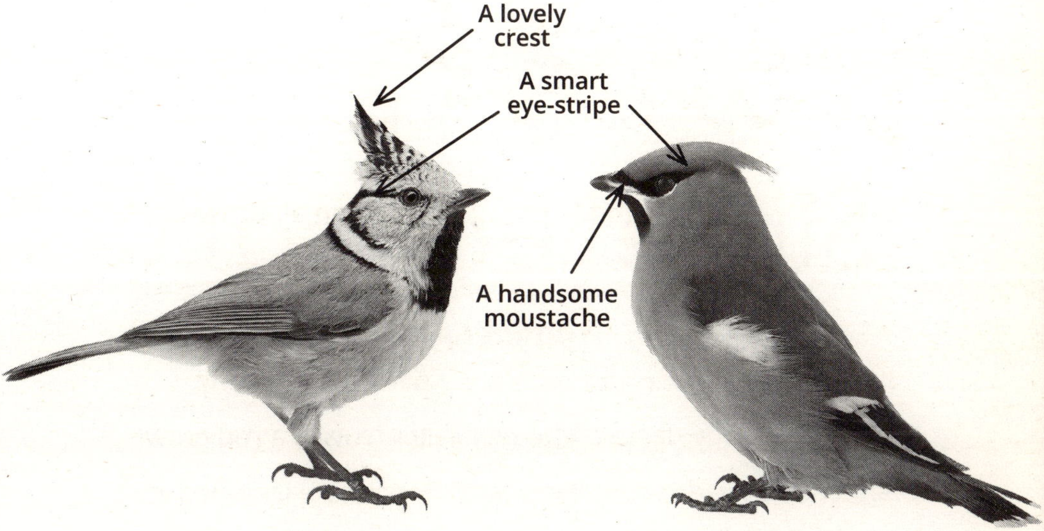

A lovely
crest

A smart
eye-stripe

A handsome
moustache

Feather colours – or **plumage** – are not just important to us. Birds use them for communication, too. Plumage tells other birds a lot of information. Straight away one bird can recognize another and can see 'that's an adult' or 'that's a female', for example.

Hamza's Nature Heroes
Sir Peter Scott
1909–1989

There are many people who inspired me with my love of nature and made me think 'I want to do that.' One of them is Sir Peter Scott. He was the ultimate bird guy in the UK.

Peter Scott was the only child of one of the world's most famous explorers – Captain Scott of the Antarctic. Wow! That really captured my imagination!

When Peter's father set off to the South Pole, an expedition that would sadly end in his death, the last thing he wrote to Peter's mother was 'Make the boy interested in natural history.' That wish definitely came true. Peter's specialist interest was birds. He was a brilliant painter who created beautiful paintings of birds, especially wetland birds. He set up Slimbridge, a wetland nature reserve in southern England. Today about 210,000 people a year visit and enjoy the wildlife there. Wetland birding is one of my particular favourites, so he was a man after my own heart.

Peter wanted to conserve wildlife, too. He understood how we are part of the natural world and that we should love and look after it. He helped set up the World Wide Fund for Nature and designed its famous panda logo, as well as the swan logo of the Wildfowl and Wetlands Trust. He saved the Hawaiian goose, or nene, from extinction.

I've been lucky enough to visit his famous study, which is a bit like my living room but way cooler. The whooper swans and Bewick's swans are within touching distance outside, and one of his final paintings is there. To me it's more exciting than going to the Oval Office or 10 Downing Street. It's an amazing place.

Sir Peter presented the BBC's first-ever natural history programme in 1953. He is one of Sir David Attenborough's heroes, which makes him a hero of mine too.

Invertebrates

There is a category of animals that has far more animals in it than the vertebrates. These are the **invertebrates** – animals with no backbone. Many of them have their skeletons on the outside of their bodies and these are called **exoskeletons**.

I move around a lot at night.

Limpet

Body: soft and squishy, the body is protected by a cone-shaped shell

Special features: a limpet's grinding teeth are made from one of the strongest materials in the animal kingdom

Relatives: scallop, oyster, mussel, clam, slug, octopus, squid

Sea anemone

Body: no head, only one opening for food and waste, stinging tentacles

Special features: some are bioluminescent (creating their own glow-in-the-dark light)

Relatives: jellyfish, coral, hydras

Earthworm

Body: long, thin body made of many segments, no legs, five pairs of hearts

Special features: earthworms can clean up land that has been polluted

Relatives: ragworm, lugworm, leech

Sea urchin

Body: soft body with a skeleton, called a test, made of many chalky plates. Tough, slender spikes grow out of the test

Special features: some have poison glands to deliver venom

Relatives: sea star, sea cucumber, brittlestar, sand dollar

Crab

Body: soft body with a hard outer skeleton, five pairs of jointed legs

Special features: can make a new, bigger shell several times as they grow

Relatives: lobster, crayfish, prawn, sea slater, water flea, woodlouse, barnacle

Spider

Body: soft body with a hard outer skin, body in two parts, four pairs of jointed legs

Special features: can make silk and has a venomous bite

Relatives: scorpion, tick, mite, harvestman

Butterfly

Body: soft body with a hard outer skin, body in three parts, called the head, thorax and abdomen, three pairs of jointed legs, antennae (feelers), has wings

Special features: butterflies taste with their feet

Relatives: fly, dragonfly, tiger beetle, ladybird, moth, glow-worm, grasshopper, earwig, flea, bee, wasp, ant, termite, water boatman

Invertebrates also include jellyfish, other types of worms, sponges, centipedes and millipedes.

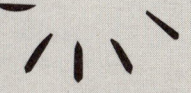

Why minibeasts rule the world!

We sometimes use the name 'minibeasts' instead of invertebrates because it's an easier word. They're also sometimes called 'creepy-crawlies' because they are small, creeping, crawling, wriggling animals. They also jump, run, fly, swim, burrow and dance. They may be small, but their importance is **enormous**.

- ⊚ Invertebrates may have been on Earth for almost 900 million years. Vertebrates have only been around for about half that time.

- ⊚ Many insects, spiders and slugs are pollinators, which means they help plants to make seeds – 160,000 of the world's pollinators are butterflies and moths.

- ⊚ If there were no bees, about 70 of the 100 crops that feed almost everyone in the world would disappear.

- ⊚ If you could weigh all the animals on Earth together,

you would find that the invertebrates make up around two-thirds of the total. That's even though they're small. They are food for other animals.

◎ Many invertebrates are decomposers. They break down dead and decaying material, such as dead leaves, dead animals, animal poo and rotten fruit, and release all that goodness back into the soil.

◎ Most spiders and many insects make silk. This incredible material is strong and versatile, used for transport, building nests and shelters, trapping prey and more. Weaver ants stitch leaves together using the silk made by their ant larvae (young).

◎ The most dangerous animal in the world (to humans, and apart from other humans) is an invertebrate – the mosquito. In some parts of the world the mosquito carries deadly diseases. And they're a menace everywhere. Even in the Arctic, you wouldn't believe how many mosquitos there are – they're all over the place, and big. You can't sit still – the mozzies hammer you day and night, and you can't go anywhere without a mosquito net.

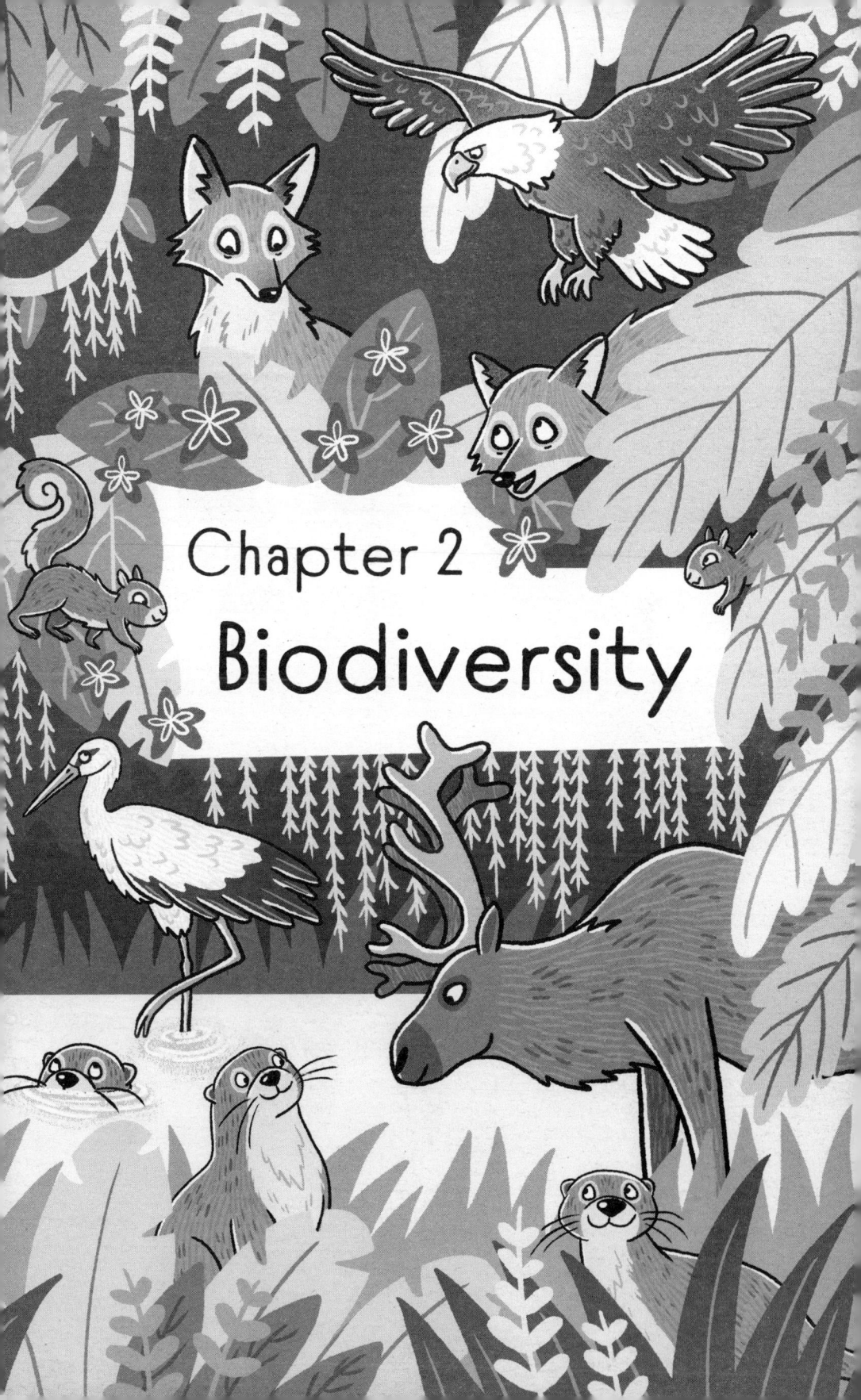

Chapter 2
Biodiversity

Hamza's Top Ten favourite animals 9

Now let's get back to my top ten countdown. In at number nine is the white-tailed eagle. It's another bird of prey – I love birds of prey!

White-tailed eagles are colossal. They are nicknamed 'barn doors' because that's what they look like in the sky – with a wingspan of up to 240 centimetres and wide, rectangular wings, they are the same size and shape as barn doors.

They are so big that when they are standing hunched on the ground they look like people, and if a pair are standing together they look like an old couple! I call my pair Lawrence and Agatha – old-fashioned names because they are old characters. One day I was giving a tour and Lawrence and Agatha were there on the horizon. I pointed them out to the group, but one lady just couldn't see them – it turned out she thought they were people sitting down!

Lawrence and Agatha are a lovey-dovey, beautiful pair. I've been watching and monitoring them for nine years, since they first got together. I know their quirks and tricks and the tag numbers on their feet.

I honestly think white-tails should be called vultures, not eagles. Most of the time they are scavenging, not hunting. They'll feed on carrion (already-dead meat), like vultures do. Do you know the scene from *Ice Age: The Meltdown* where the vultures are singing a song while they're waiting for the animals to die? Or *The Lion King* where the vultures are gathering round an unconscious Simba? That's what the white-tails are like. Sometimes I see them just sitting, watching otters catch fish. The white-tails hassle the otters or distract them until they drop the fish – an easy meal for an eagle. So sometimes white-tails end up eating deep-water fish such as wrasse that they could never catch for themselves.

I can be a talon-ted hunter!

That's not to say that they can't hunt if they need to. I was lucky enough to film white-tails hunting barnacle geese on the Isle of Islay for *Wild Isles*. Thousands of barnacle geese had flocked together for safety, but young white-tailed eagles singled out an exhausted goose that didn't manage to keep up with the others. They worked together to chase the goose, with one of them finally catching it with its sharp talons in mid-air and using all its strength to avoid dropping its prey.

White-tails play an important role in the environment and are a sign the area is healthy.

Facts

Scientific name:	*Haliaeetus albicilla*
Bird family:	hawks, eagles, kites and vultures
Height:	70–90 cm
Wingspan:	180–240 cm
Found:	rocky sea coasts, islands, large lakes
Eats:	fish, gulls, ducks, rabbits, hares, carrion
My three words:	colossal, magnificent, soaring

White-tailed eagles often nest in larch trees. We don't know for sure why they do this. They probably choose them by sight, as larch trees are easy to recognize and often taller than surrounding trees. Larch and pine wood have antibacterial properties, which means they contain chemicals which stop bacteria from growing. This is good news for keeping the eagle's nest clean and disease-free.

The white-tailed eagle appears on Germany's coat of arms.

They are speedy in flight and can reach 70 kilometres per hour.

They don't have any natural predators.

Habitats and biomes

You probably know that the place where an animal or plant lives is called a habitat. This is a place where it can find food, shelter and a safe place to breed.

Some animals can only live in a very particular habitat. One example is the Kirtland's warbler, a North American bird that only nests in young jack pine trees, which in turn only grow after wildfires.

Others can live in almost any habitat and across great stretches of the world. The red fox, brown rat and housefly are top examples. The red fox is found in 83 countries and five continents, in forests, grassland, farmland, mountains, deserts and even city centres.

A habitat can be as small as an individual pond or tree or as big as a vast rainforest. But habitats fall into some broad categories called **biomes**.

There are four major biomes on land:

- **Tundra** – in the far north and cold, or even frozen, all year round, with mosses, lichens and grasses.

- **Forest** – full of trees. Includes taiga – conifer forest (mostly evergreen trees with cones and needles) just south of tundra; deciduous forests or woodlands, which are warm and mild with noticeable seasons; tropical rainforests, which are are near the equator and are hot and wet all year round.

- **Grasslands** – vast, open areas of land where grasses are the main plants. Savanna is a type of tropical grassland where it's hot all year round and there are wet and dry seasons.

- **Deserts** – dry all year round, can be hot or cold.

Ocean biomes include coral reefs, estuaries and the open ocean. But they also have different zones – from the sunlight zone in the top 200 metres near the surface, through pelagic and benthic zones to the deepest trenches below 6,000 metres. That makes lots of different habitat possibilities.

55

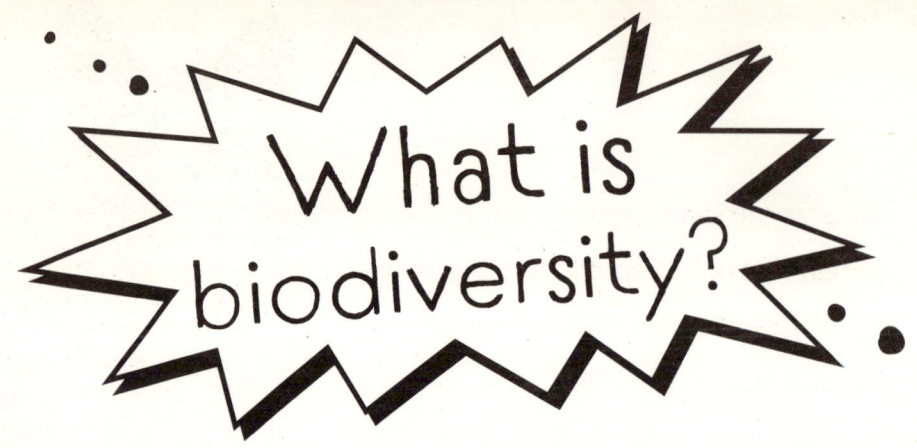

What is biodiversity?

Walk through your local park and you may see squirrels rushing about, blackbirds and robins in the trees, perhaps pigeons, mallards and even a jay. Above you there are different kinds of trees. Get down to ground level and you may see ladybirds, worms and beetles, and in summer there may be butterflies fluttering from flower to flower. There are probably many more that are hiding or come out at night.

This variety is called biodiversity. It means the variety of animals, plants and fungi. We talk about the biodiversity of a particular habitat, region or of the whole Earth.

Biodiversity is important because the world is a richer place with it. Living things are connected to each other in intricate ways, with one species depending on others for food and each having an effect on the environment. We'll see this later in the book at Yellowstone National Park in the USA, where wolves had disappeared and this had set off other changes in the environment and its inhabitants. Bringing wolves back helped restore the landscape. If numbers of one living thing rise and fall, others can usually cope and adjust, but if they disappear altogether, who knows what knock-on effects may occur. We don't even know about all the connections in nature.

We humans need biodiversity in so many ways. We need pollinating insects so we can grow our food crops. Deep in rainforests we have have found plants that give us life-saving medicines, and there may be more to discover.

Biodiversity is under threat. As of 2023, the UK has lost around half its biodiversity compared with 200 years ago – this is the worst of any of the world's richest countries. Many organizations are working hard to improve this.

Find out more: Yellowstone wolves, page 218; rewilding, page 78; beaver reintroduction, page 66

Hamza's habitats
Scots pine ancient forest

Habitat: Scots pine ancient forest

Where in the world: a few remote, mountainous areas of Scotland

Landscape: dense forest

If you are ever lost in a Scots pine forest and need to find your bearings, look for a wood ant nest. It will always be on the southern side of a tree, where it's warmer.

One of my favourite habitats is the ancient Scots pine forests of Scotland. This rich, ancient habitat dates back over 9,000 years with many individual trees reaching between 400 and 500 years old.

The dominant tree is the Scots pine. It's a magnificent tree – a conifer (which means its seeds develop in cones) – and it is evergreen with tough needles. It creates a brilliant habitat for hundreds of other species, from the lichens that live on its bark to the wood ants that make enormous nests at its base. Squirrels and pine martens dart through its branches and birds feed and nest in its canopy. A golden eagle's favourite nesting spot is in Scots pine trees (although it prefers open areas to forest).

Star animals of the ancient Scots pine forest include the capercaillie, the extremely rare Scottish wildcat, red squirrel, crested tit and crossbill.

Once there were wild boar, lynx, grey wolves, elk and beavers. A pair of beavers was reintroduced in 2023 and perhaps other ancient species will be brought back one day. I would love to see lynx stalking through the trees.

Spotlight on

Hedges

A habitat has to provide certain things for the wildlife it supports: food, water, shelter and space. A hedge does all of this. So many different animals live in one hedge, at different levels and heights, that I call hedges blocks of flats for wildlife! And the block of flats has enough shelter and space for all sorts of animals . . .

Beetles munch away at the hedge's leaves and **bees** slurp energy-rich nectar from its flowers.

◉ **Millipedes, snails and worms** crawl underneath, feeding on dead leaves and helping recycle the leaves' goodness.

◉ **Small mammals** such as **mice** move up and down eating seeds, berries and insects. **Hedgehogs** may snuffle underneath, looking for worms, beetles and birds' eggs.

◎ **Birds** such as sparrows, robins, wrens and finches forage for food and also make their nests in the hedge. Thick branches and thorns make it a good, safe place. Birds also perch on top to sing, keep watch over their territory or to look out for danger.

◎ **Foxes and deer** may hide or hunt for a meal in a hedge, **crows** may raid the birds' nests for eggs, and **birds of prey** know that a good moment to catch small animals is just as they leave the hedge or are about to go back in.

◎ **Flowers** grow in the shelter of the hedge, protected from the wind and benefiting from the nutrients it adds to the soil.

A hedge is one of the best things you can possibly have in your garden.

A hedge is a wildlife corridor as well as a block of flats. Unlike walls, fences or roads, it doesn't make a barrier but allows the animals to pass through as they travel from place to place. The corridor links patches of different habitat: it's a mini strip of woodland between a wood and a park, a haven between two bare fields or a hiding place alongside a river.

It's a phenomenal habitat. When you lay a hedge you're doing the same thing as a beaver does, coppicing the wood and making a new environment. Over the last 50 years, we have lost hundreds of miles of hedges, but now people are rediscovering how important they are and learning to love and care for them again!

Biodiversity in the UK

There are roughly 86,000 different species in the UK. The pie chart shows you which groups they all belong to.

Sadly, 16 per cent of Britain's animals and plants are at risk of extinction.

The Scottish crossbill, Skomer vole, Shetland wren and the Scottish wildcat are some of the animals that are only found in the UK and nowhere else in the world.

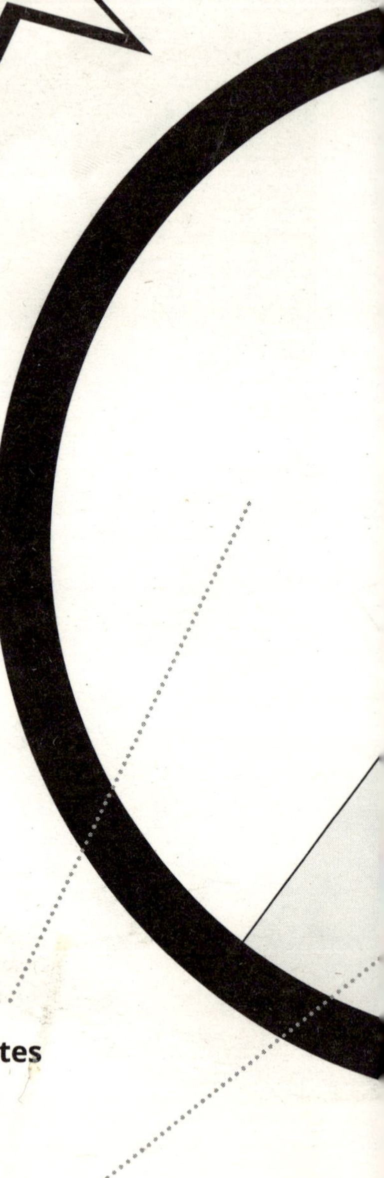

Other Invertebrates (39,950)

Insects (24,000)

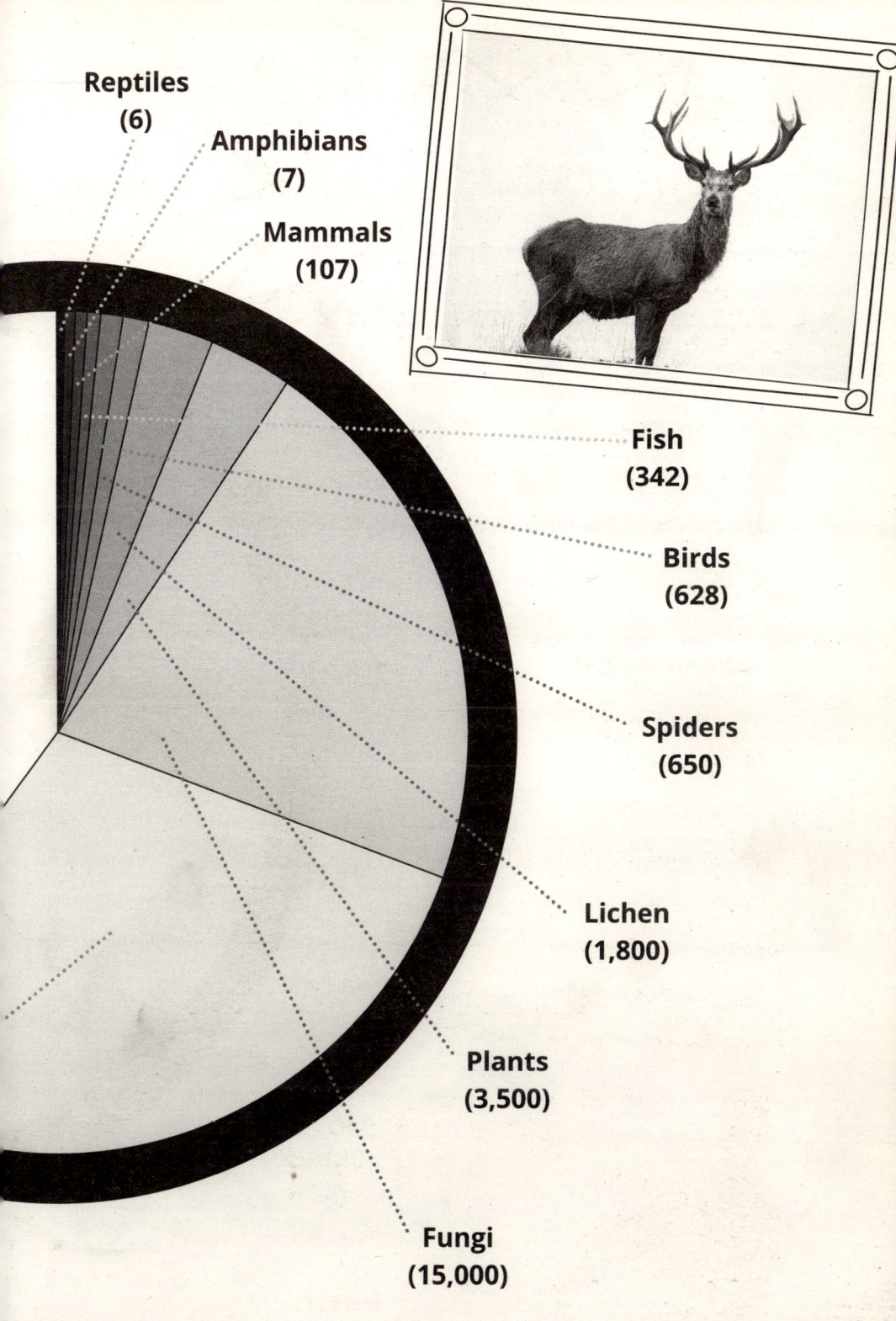

Reptiles
(6)

Amphibians
(7)

Mammals
(107)

Fish
(342)

Birds
(628)

Spiders
(650)

Lichen
(1,800)

Plants
(3,500)

Fungi
(15,000)

The return of the beaver

Sometimes, if I'm out and about somewhere quiet on the east coast of Scotland, just before dusk, I might be lucky enough to get a glimpse of a beaver.

These stealthy swimmers went extinct in the UK in the sixteenth century. They were hunted for their fur and because people wanted to drain land for farming.

Their absence made the environment poorer. Beavers are ecosystem engineers and keystone species, which means they shape and create habitat and affect the

wildlife in it. They use their strong, large, continuously growing front teeth as tools to cut down tree trunks and branches to strip the bark. They eat the bark and soft wood, and they need to fell the trees because they can't climb up to eat the soft new twigs. They use the branches to build a dam across a river.

Behind the dam, the beaver builds its home from branches and sticks. It's called a lodge. Beavers have their babies (kits) here.

Find out more: Keystone species, page 76; ecosystem engineers, page 74

The dam and lodge act like a sponge, slowing the flow of the river, and shallow pools form, which are great for all sorts of wildlife. Moving all that wood is heavy work, so the beavers also dig little canals so that they can float the wood to where they need it.

Beavers chop down trees, but this allows new trees to grow. They change the way that water flows in a landscape. They create flooding in some areas, which can make them unpopular with farmers. However, they can reduce severe flooding events. If water has little pools and beaver-made canals to spread out into when it rains heavily, that water will flow slowly through the landscape rather than rushing across the land and causing floods.

The first official UK beaver reintroduction happened in 2003, when conservationists released two beaver families from Norway at a reserve in Kent. Today there are beaver reintroduction projects all over the UK, with approximately 2,000 beavers now living here.

Nature jobs

Wildlife reserves are important places where we protect habitats and the plants and animals that live there. There are lots of different jobs for nature lovers in wildlife reserves.

I'm a tree officer – I plant trees and take care of them. Sometimes I have volunteers to help me.

I pick up rubbish that people have left behind or which has got into the rivers and streams. This helps make the habitat healthier for wildlife.

I'm a researcher, studying the wildlife that lives here, and today I'm identifying invertebrates in the river.

I'm a hydrologist, which means I monitor the water. I measure how much of it there is, how it flows, and test how clean it is.

I'm a campaigner. I tell people what the nature reserve is doing so that they know how important it is. This might persuade organizations or governments to give us money to do our work.

I work on events – today it's a pond dipping session for local schoolchildren.

I'm building an accessible path so that everybody can enjoy the reserve.

Ask at your local wildlife reserve to find out what activities you can get involved with! Maybe there will be a minibeast hunt, a kids' photography day, a guided birdwatching session, pond dipping or even a bat safari.

World biodiversity hotspots

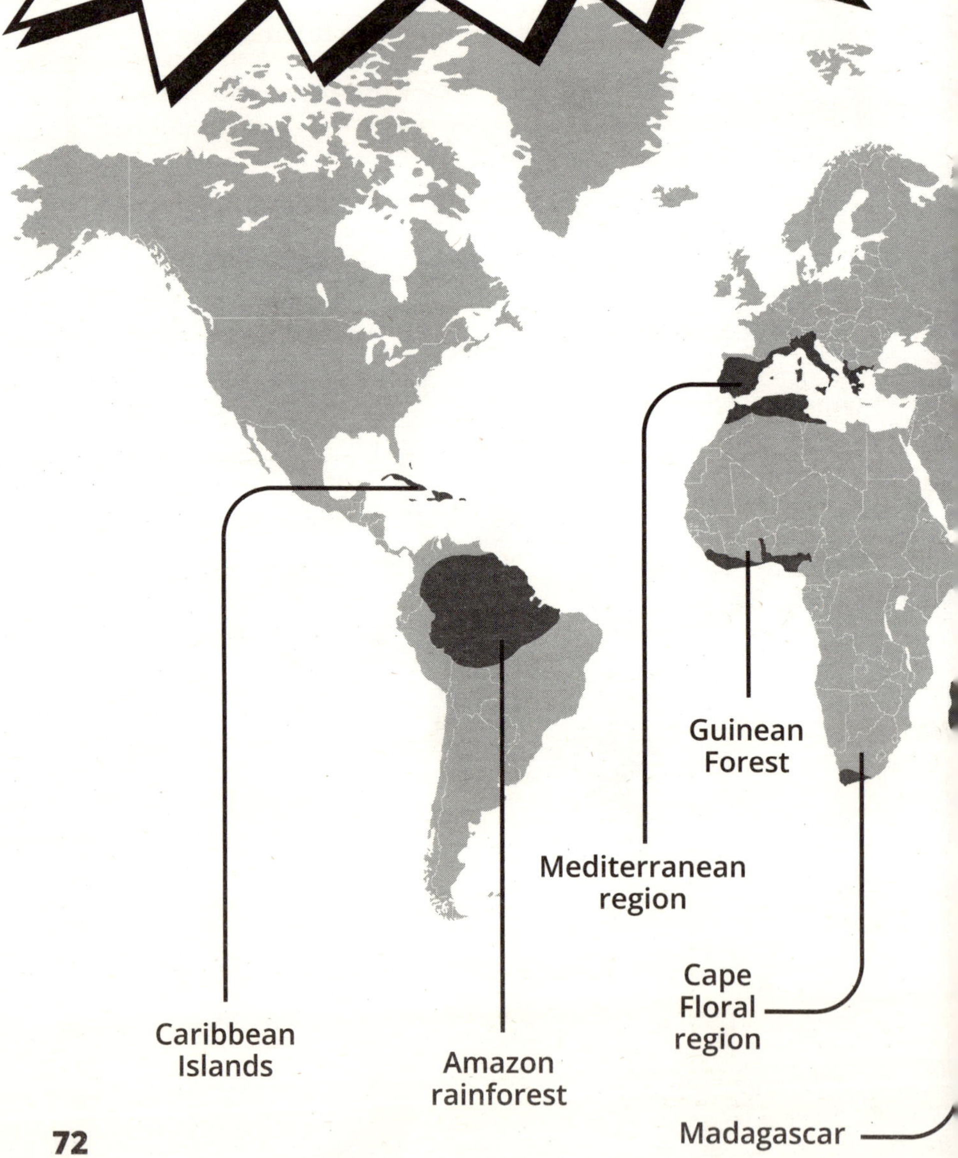

Guinean Forest

Mediterranean region

Cape Floral region

Caribbean Islands

Amazon rainforest

Madagascar

Here are ten of the world's most incredible biodiversity hotspots. These are places with particularly rich biodiversity that are also threatened by habitat loss and human activity.

The Amazon rainforest is home to more than **3 million** different species of living things.

The Highlands of Scotland – not far from where I live – is the most biodiverse region in the UK. But it does depend on what you're looking at. The most biodiverse for mammals, with 44 species found in a recent study, is the little Isle of Purbeck in Dorset.

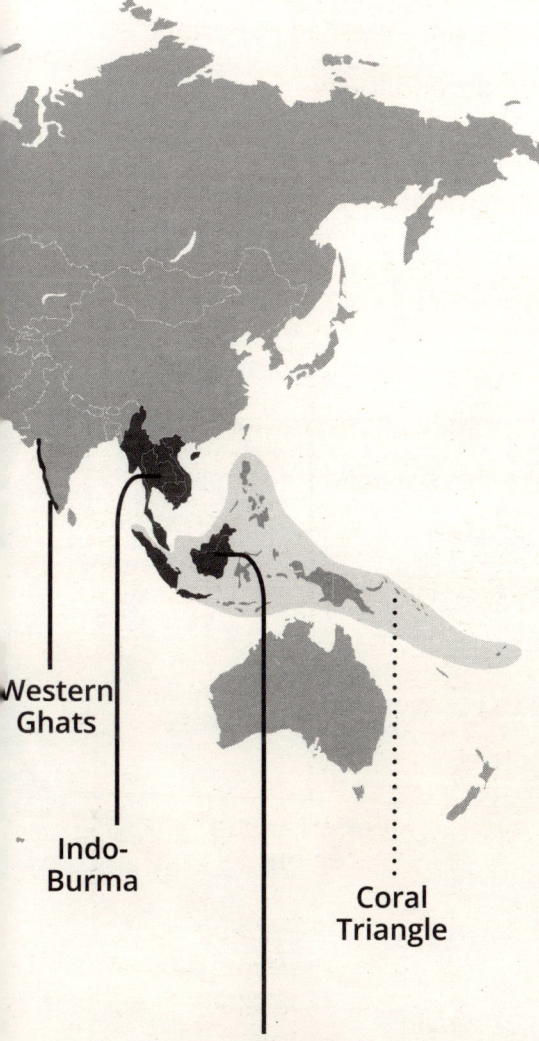

Western Ghats

Indo-Burma

Coral Triangle

Sundaland

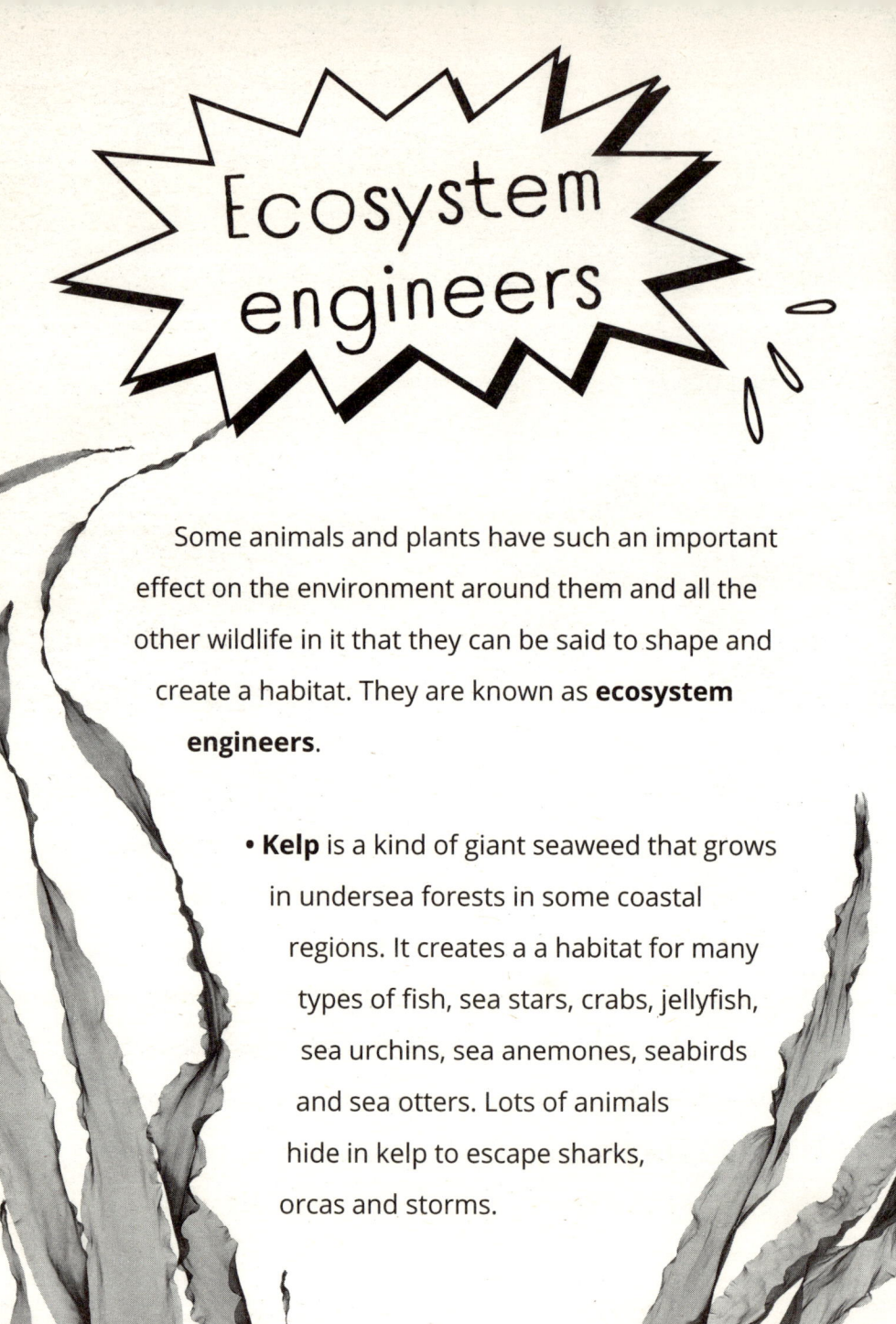

Ecosystem engineers

Some animals and plants have such an important effect on the environment around them and all the other wildlife in it that they can be said to shape and create a habitat. They are known as **ecosystem engineers**.

• **Kelp** is a kind of giant seaweed that grows in undersea forests in some coastal regions. It creates a a habitat for many types of fish, sea stars, crabs, jellyfish, sea urchins, sea anemones, seabirds and sea otters. Lots of animals hide in kelp to escape sharks, orcas and storms.

- **Reef corals** build a coral reef, a living structure that provides food, home and shelter for millions of other living things.

- **Elephants** change the environment in various ways. Their migration routes carve trails in the land, which fill with rainwater to create small ponds. They push trees over and eat vegetation, sometimes clearing large areas of forest and creating grassland. They spread plants' seeds over vast distances. They even dig wells and create watering holes.

Keystone species

Some animals are so important to an ecosystem that without them the whole ecosystem starts to collapse. They may not actually shape the environment like an ecosystem engineer does, but without them there is a knock-on effect that leads to loss of biodiversity. These are known as keystone species.

• For example, **sea stars** do not create a rock pool environment, but if they disappear, mussels and barnacles – which they eat – increase so much that nearly all the other creatures in the rock pool may disappear, too.

• The **wolves** of Yellowstone are another example of a keystone species.

• **Sea otters** in kelp forests, keep the numbers of sea

urchins down. Kelp forests in northwestern Canada were disappearing in the early 2000s and people couldn't figure out why. It turned out that sea urchins were eating all the kelp. Few animals were eating the sea urchins because numbers of sea otters had declined. Helping the sea otter population to recover brought down the numbers of sea urchins and that increased the number of the kelp forests, bringing everything back into balance.

- **Beavers** are both ecosystem engineers *and* a keystone species.

Find out more: Yellowstone wolves, page 218

Rewilding

One way to tackle biodiversity loss is to help places grow wild again. This is called rewilding. It's a way of making some changes to the landscape and then allowing nature to do its own thing.

A famous rewilding site in the UK is **Knepp** in Sussex, England. Knepp's owners Isabella Tree and Charlie Burrell started this project in 2000 on land that had been poor farmland. To start with they removed most of the farm's fences and stopped trimming the hedges. They introduced free-roaming cattle, pigs and ponies that would live wild lives, and added fallow deer to the red deer and roe deer already living there. All of these herbivores eat different things, so have different effects

on the landscape and wildlife. The owners removed weirs and canals so that the river would start to follow a natural wiggly course and wetland areas would reappear.

At Knepp there have been reintroduction success stories – the white stork and the beaver. But this is not really what Knepp is about. The rewilding has brought all sorts of wildlife that has appeared of its own accord – much of it a surprise. The site now has the country's largest populations of purple emperor butterflies and turtle doves, and is a hotspot for nightingales.

In Scotland, there are several rewilding projects. **Cairngorms Connect** aims to restore the different habitats of the Cairngorms, including Scots pine ancient forest, mountain peaks, peat bogs and the River Spey floodplain. Many habitats have been taken over by too many deer and too many conifers (grown for timber), so people are working to restore these habitats so they can support more animals and plants.

In my own local area, the **Kilchoan Estate** rewilding project involves both land and sea. On land some of the sheep pasture is being transformed into a patchwork of different wild habitats. An oyster nursery has been set up at sea to help restore the population of the native oyster, an important ecosystem engineer that builds reefs and cleans the water. In the local community people avoid using pesticides, which is why I can see so many barn owls! The area is home to the flapper skate, the world's biggest skate (a relative of sharks). It's a huge, 3-metre-long fish and the females travel to these waters to lay their eggs.

It's also important to help connect people with nature so that everyone understands how important the natural world is, and everyone can enjoy it and protect it.

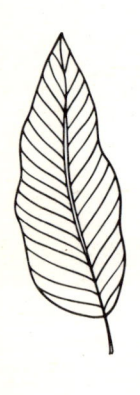

One of the most extreme examples of rewilding is taking place in **Chernobyl** in northern Ukraine and in Belarus. In 1986 there was a terrible disaster when a nuclear reactor at Chernobyl exploded. An exclusion zone covering an area of around 2,634 square kilometres (more than twice the size of London) was created around the site, which is too dangerous for people because of the harmful radiation released by the explosion. Over 100,000 people had to leave their homes and since then people can only visit under strict conditions and with protective gear.

An unexpected result of there being no people is that nature has returned. There are hundreds of plants and animal species in the exclusion zone. Vegetation has invaded the city ruins. Lynx, bison, wild boar, elk, wild horses and even wolves roam the streets and the nearby pine forests. Black grouse perform their lek, a kind of courtship display. Eagles raise chicks in blocks of flats. There are no humans for them to fear. Wildlife has proved to be remarkably resilient.

Rewilding is also helping to tackle the climate crisis. Areas with high biodiversity hold on to more carbon compared with other areas. Healthy water systems, with soil that holds water well, reduce the risk of wildfires and floods. Rewilding principles can even be applied to farmland so that food is produced in a way that works in closer harmony with nature.

Chapter 3
Sensing

Hamza's Top Ten favourite animals

8

In at number eight is – you guessed it – another bird of prey. It's my absolute favourite bird of prey: the **golden eagle**. It was these eagles that first drew me to Scotland. I just had to move there, whatever it took, to find them.

The golden eagle – or goldie – is the king of the sky. No, more than that, it's the emperor. It rules. It's not the biggest of our eagles; the best way I can describe it is to imagine a big heavyweight boxer – like Anthony Joshua. That's the white-tailed eagle, our biggest eagle. Then imagine a featherweight boxer, not as big but nifty, speedy and brave. He throws his punches and runs around, and he might get caught once or twice by the heavyweight but he's got fitness and agility and he can really go for it. That's the goldie.

The golden eagle is the top bird of prey, the supreme hunter. Assassin. The pair I watch, I call them Lady because she's the Lady of the Sky and Assassin because one day I actually saw him bring a full-grown fox and carry it to the nest for his chicks.

One pair of goldies that I've filmed nested in an old, tall Scots pine tree called a 'granny tree'. The Scots pine is the largest and longest-lived native tree in Scottish forests. Individual trees can grow up to 35 metres high and they provide a home or food for many different species of wildlife.

When I was filming for *Wild Isles*, I was in the Cairngorm mountains and I had the task of filming the adults and chicks. I got some great footage of the adults being all gorgeous, but I was called away to another job and I didn't manage to film the chicks. My good friend and one of my mentors John Aitcheson captured some of the most incredible footage of golden eagle chicks I have ever seen. He filmed them climbing down the tree, walking around the heather like big turkeys, and then in the evening going back up the tree to the nest, ready to be fed by the parents. No one had ever filmed anything like that before. I wanted to go back to film something similar for myself.

When I went back the next year to film it, the same nest had failed, but my friends Ewan and Jenny, who are eagle experts, took me to see another golden eagle nest. They had special permission to climb up to the nest to study it. The nest was in an old Scots pine that had had golden eagles living in it for over a hundred years.

Ewan climbed up to the top and saw there were two chicks there! Carefully he lowered them down in a bag down to Jenny and me. Now it was time to meet them! I couldn't believe it. I was so excited. We ringed the chicks, which means putting tags like bracelets around their legs. We measured the talons of one of them as 14.8 cm – big feet, which tells us that it was a female. Then Ewan climbed back up to the nest and put the chicks back. I was covered in little fluffy eagle-down feathers.

Female birds of prey are generally bigger than males, which was why the big feet meant that we knew that our chick was female.

Hopefully I'll see these same chicks again in the sky over the next few years.

87

Facts

Scientific name:	*Aquila chrysaetos*
Bird family:	hawks, eagles, kites and vultures
Height:	75–88 cm
Wingspan:	190–230 cm
Found in:	open countryside, moorland, mountains
Eats:	birds, mammals (from mice to voles to otters and foxes), carrion
Eggs:	1–3 per clutch
My three words:	assassin, regal, ruler

Eagles, like all birds of prey, have incredibly sharp eyesight. An eagle can see something from 6 metres away that you or I can only see from 1.5 metres away. They can see a hare move from 2 kilometres away.

Birds of prey can also see more colours than we can. And they can see ultraviolet light.

Golden eagles can focus on two things at once: they can see forward and to the side at the same time, which helps them look for prey.

They can fly at great speed, which means the world rushes past at great speed, too. Their sight needs to be sharp enough that they can process this high-speed information.

Animal eyes

Eagles have some of the sharpest sight in the animal kingdom. It's part of what makes them such effective predators. All animals' eyes need to be just right for the lives they live.

Forward or all-around

Usually, predators have their eyes close together at the front of their head, facing forward. This is called **binocular vision**. The portion of the world that can be seen by each eye overlaps. This gives a sharper view.

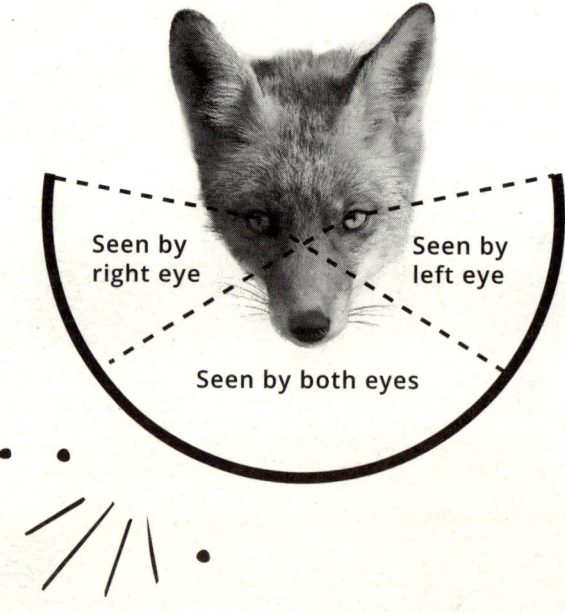

Seen by right eye

Seen by left eye

Seen by both eyes

Usually, prey animals have their eyes spaced wide apart, on each side of their head. This is called **monocular vision**. Each eye is looking at a different place, which means the animal gets a much wider view around them.

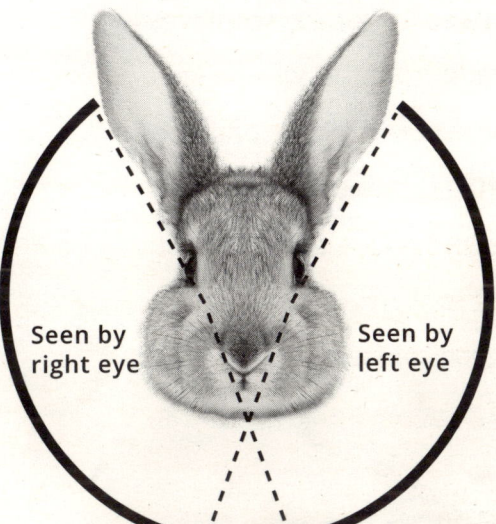

Seen by right eye

Seen by left eye

Seen by both eyes

Chameleons can move each eye **independently** of each other, swivelling each one to look in all directions. This gives them almost an all-round view.

Seeing in the dark

Some animals have excellent night vision. They can see in very low light. Owls, tarsiers and bush babies all have very big eyes for their size and at the back of the eye is a highly reflective layer (like a hi-vis jacket!). A tarsier's eye is as big as its brain!

The world in colour

How many colours can you see? Hundreds? Thousands? Humans can see light in what's known as the visible spectrum. In our eyes we have cells that are sensitive to red, green and blue light, and from this we can see a whole range of colours. Some of us have a condition called colour blindness which means we can't distinguish quite as many colours, but even so it's a colourful world.

Some animals can't see all these colours. Others can see extra ones that we can't see.

Most mammals can only see in black and white or in a lesser range of colours than we can. Your dog, for example, can only see shades of blue, green and yellow (no red). Try throwing an orange ball for a dog and watch it run around looking for it! But other primates (apes, monkeys, lemurs, tarsiers and so on) can see in full colour like we can. It's thought we evolved this skill so we can tell ripe from unripe fruit.

Many animals can see **ultraviolet light** which is just beyond the human visible spectrum.

◎ Insects such as bees and butterflies can see flower patterns that we can't. Certain flowers have patterns of lines, like wheel spokes, visible in ultraviolet light, emphasizing the centre of the flower. These point out where the flower's nectar is.

◎ To us, all blue tits look pretty much the same. But blue tits themselves can see the difference. Strong, fit birds have an extra-vivid, ultraviolet sheen to their blue crests. This helps birds choose a mate that will be a successful parent.

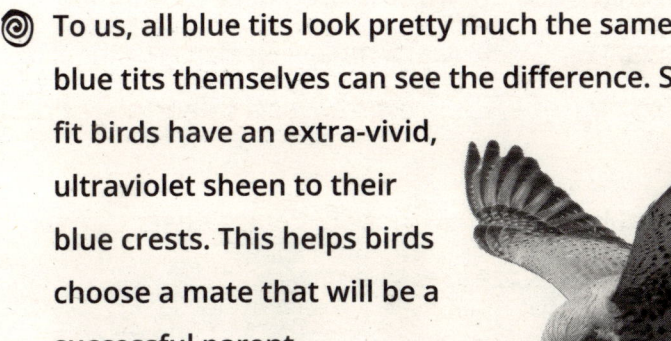

◎ Kestrels can see mouse wee glowing in ultraviolet. Mice wee constantly as they run along, so their wee leaves trails which to the kestrel glow like a neon sign. To us a kestrel looks like it's scanning a wide area looking for prey, but these track lines are helping it to zoom in on the right spot.

MOUSE

High and low sounds

200-300,000 Hz
9,000–120,000 Hz
1,000–100,000 Hz
48–64,000 Hz
40–60,000 Hz

20–20,000 Hz
675–11,500 Hz
20–4,000 Hz
50–1,000 Hz

16–12,000 Hz
7–35,000 Hz
2–160,000 Hz

Some animals can hear extremely high sounds, some hear extremely low sounds, and others hear somewhere in between. At the same time, some can hear over much longer distances than others.

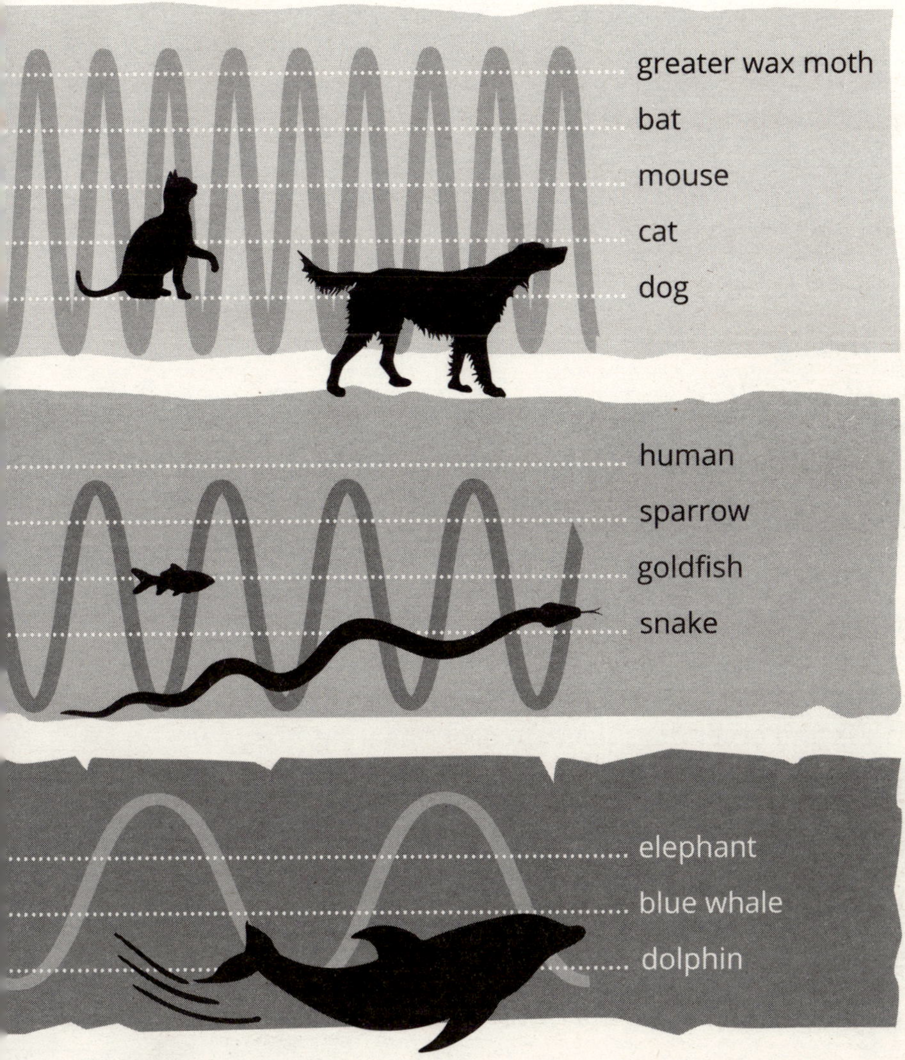

greater wax moth

bat

mouse

cat

dog

human

sparrow

goldfish

snake

elephant

blue whale

dolphin

Hz stands for Hertz and is a unit of frequency of sound waves. A high number means fast sound waves and a high sound. A low number means slow sound waves and a low sound.

Hearing

Many animals depend on hearing to find their way and locate their prey. In fact, most animals have better hearing than us humans.

Sound is a vibration that travels through air, water or the ground, and animals pick up these vibrations in different parts of their bodies:

- ◎ **Through tiny hairs, as with insects.**
- ◎ **Through their skin, as with snakes on land and fish and other sea creatures in water.**
- ◎ **And through their ears!**

Ears are shaped so that an animal can hear as well as possible.

Wolves and other dogs have excellent hearing. They can hear sounds four times as far away as a human can, and that are too high-pitched for a human to hear. They have lots of small muscles in their ears and can move them very accurately to pinpoint where a sound is coming from and how far away it is.

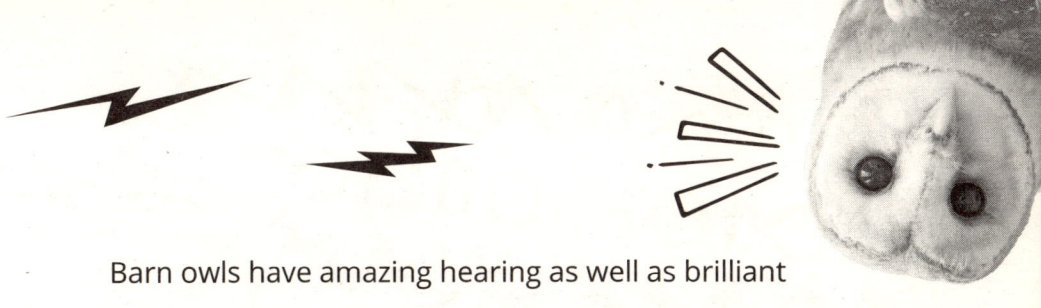

Barn owls have amazing hearing as well as brilliant eyesight. Like most birds, their ears are just holes, with no external parts. The soft feathers on an owl's face are arranged in a disc shape – rather like a satellite dish – to help funnel the sound towards their ears. They also have one ear slightly higher than the other, which helps them work out where a sound is coming from. The sound will reach one ear a millisecond quicker than the other, and their brain will process it to tell them where the sound is coming from. We do that too, just not as well as the owl does.

Elephants can hear sounds that are lower than

Can you hear me over there?

humans can hear, called infrasonic. They can call to each other over very long distances.

Pigeons can hear infrasonic sounds, too. We are not completely sure what they are hearing, but we think they might be listening for nearby storms, to avoid flying through them.

Find out more: Animal calls, page 258

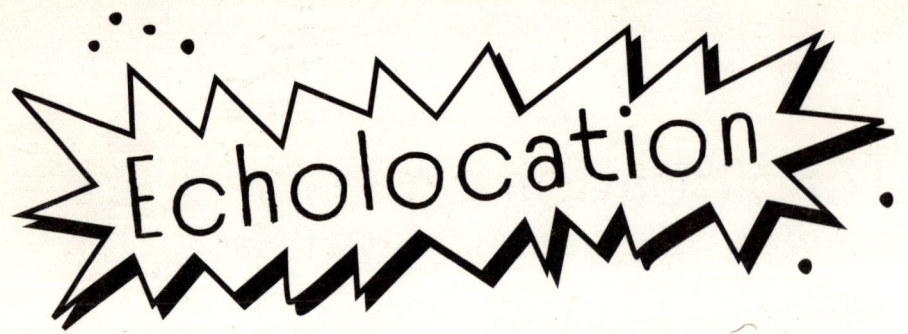

Echolocation

Echolocation is like a hearing **superpower**. It means using echoes to find things. It's a superb way of sensing things in the dark.

An animal using echolocation makes very high-pitched squeaks and calls. The sound, like any other sound, travels through the air or water. A bat's squeak, for example, hits other objects, such as tiny flying insects, other bats, or trees, walls and things that the bat doesn't want to bump into. The squeak rebounds, or echoes back, off the object to the bat's ears. It can then tell what the object is, where it is and even if it is moving.

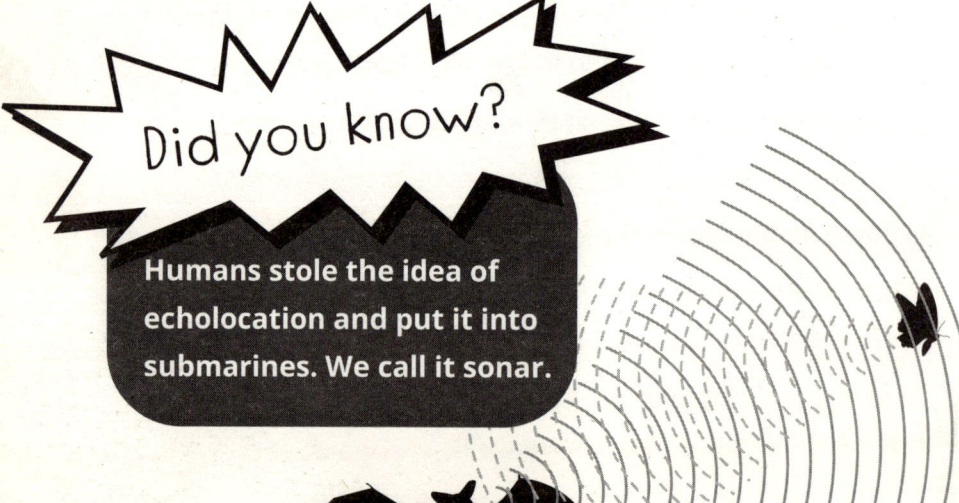

Did you know?

Humans stole the idea of echolocation and put it into submarines. We call it sonar.

Echolocation experts

Bats

Bats can catch night-flying insects. Most birds can't do this, which means all the more insects for the bats. Bats' hearing is so good that they can hear an insect 5 metres away, pinpoint exactly where it is and dodge any obstacles in its way. It turns off its own hearing for a split second before it calls, so it doesn't deafen itself.

Whales and dolphins

Echolocation is a very useful skill underwater. Whales and dolphins make a high-pitched clicking sound. The shape of their head allows them to send the sound in a clear, straight beam. They interpret the echoes to tell where an object is, how far away it is, what size it is and how fast it is moving. They can tell what it is (and whether it's something they want to eat). If you've ever seen an ultrasound picture of a baby before it's born, you will have an idea of the kind of picture whales and dolphins can build in their minds using echolocation.

Smell

Smell helps animals find food, find each other and tell when danger is coming. It's a very useful sense, because lots of things are quiet or can't be seen – but they can be smelled. Animals can also pick up and send lots of information through smell.

Smell is possible because of tiny amounts of chemicals in the air. Every single thing gives off tiny specks of itself. Flowers give off sweet-smelling chemicals. Sweat, wee and poo are stinkier but can also be a way for one animal to find another.

Animals smell things when these tiny specks reach special chemical-detection cells, called olfactory cells, in their noses. Some animals have *lots of*

Find out more: elephant senses, page 110; memory, page 394

these chemical detectors. A brown bear's sense of smell is 2,100 times better than ours! But elephants are the champion smellers. They can smell water underground from a long way away and also remember smells for many years.

Otters can smell underwater. They blow out a bubble, and as it touches a fish or other object, they breathe it back in and pick up the smell.

Sharks are also expert smellers. Dolphins, on the other hand, have no sense of smell at all.

Is that dinner I can smell?

On the scent

When animals are trying to find prey by smell, wind is the answer. That's not a joke! Animals use the breeze and air currents to help them work out where a smell is coming from and follow its path. You have probably seen a dog sniffing along an invisible path on the ground.

When a predator is near its prey, it wants to make sure its own scent doesn't give it away, so it keeps downwind of the prey (which means the wind is not blowing the predator's scent towards the prey).

I use this skill when I'm filming. Once, I spent weeks preparing to hide from pine martens by trying to make them used to my smell. For those weeks I would wear a T-shirt for a few days and then leave it somewhere in the forest where I was going to film. I would make sure I used the same soap and shampoo so that I always smelled the same. After a few days, lots of my T-shirts were scattered around the place!

I was making the smell of me part of the 'smell landscape' of the forest so that when I turned up with my camera to wait for the pine martens they wouldn't get upset. Pine martens are predators that are mainly active at night, so it's no surprise that they have a very keen sense of smell.

If you're trying to find or photograph elusive wildlife, it's a good idea to remember about smell! Here is one of the pine marten pictures I managed to get:

10 Smelly Animals

Birds

❶ Hoatzin or stinkbird

Smells like manure because of its unusual digestive system

❷ Wood hoopoe

Squirts a liquid that smells of rotten eggs at its enemies

❸ Fulmar chicks

Squirt smelly, poisonous sick at intruders that come near the nest

Invertebrates

❹ Stink bug

Releases a smelly chemical through pores in its body to defend itself against birds and lizards

❺ Millipede

Covers its body with a poisonous and very smelly fluid

❻ Bombardier beetle

Squirts smelly, burning acid from its bottom to fight off attackers

> **I've been squirted with sick by a fulmar chick – yuck!**

Mammals

7 Skunk

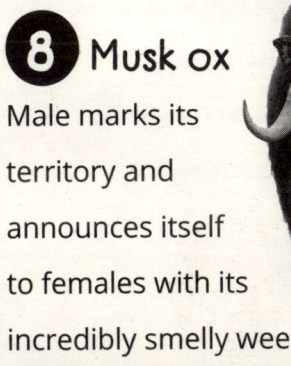

Squirts smelly liquid from its bottom when it's in danger

8 Musk ox

Male marks its territory and announces itself to females with its incredibly smelly wee

9 Elephant

A bull during mating season pongs for miles around

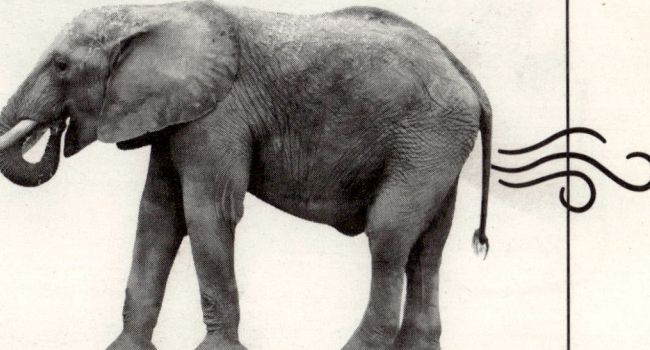

Reptiles

10 King ratsnake

Squirts smelly liquid from – you guessed it – its bottom

Taste

Taste is a similar sense to smell. It's another way of picking up those tiny specks of chemicals.

Most animals use their **tongues** to taste with, and it's the taste buds that do the job. Humans are pretty good at tasting – we have 9,000 taste buds, whereas a cat only has 400. Cows have over 25,000!

Some animals taste with other parts of their body. Butterflies taste with their feet, octopuses taste with their arms and the catfish has taste buds all over its body. It may have as many as 250,000 taste buds covering its skin!

Taste is useful because it allows animals to choose the right food for them.

Yum, grass!

Animals that eat meat tend to like meaty, savoury tastes. Grass-eating animals like bitter tastes and animals that eat fruit like sweet tastes. Birds can't taste chilli – or hot food – at all. Taste is also a warning if something is bad to eat – like the sour taste of 'off' milk.

Some animals use taste as a weapon. Poison dart frogs have poisonous skin that will kill any predator that tries to eat them. Predators learn to recognize the nasty taste and avoid eating poison dart frogs in future. The poison dart frogs get their poison from chemicals in the ants they eat. But poison dart frogs in zoos are fed on different kinds of ants and don't make the poison at all.

POISON

PREDATORS BEWARE

Touch

Our **skin** is a sensitive organ that allows us to touch and feel the world around us. We can find out quite a lot just by touch, but our sense of touch is not as good as that of some of these animals . . .

The **star-nosed mole** uses its amazing star-shaped nose to feel its way underground, and it does so really fast. All moles find their way using smell and touch as they can't see very well.

Many wading birds poke their **bills** into mud and feel vibrations to find small creatures wriggling under the surface. Snipes are the experts at this. Kiwis do this too.

Most mammals and some birds have **whiskers**. These are fantastic tools for sensing through touch. Cats, including big cats, for example, rely on their whiskers. Their whiskers are as wide as the widest part of their body so they are perfect for feeling the size of something. If the cat wants to know if it can fit through a gap, it just needs to try its head first. If it feels its whiskers touch the sides, then it knows the answer is no. But this isn't all. Through their whiskers cats can feel wind currents and even the movements of their prey.

Spotlight on

Elephant – super-senses

Elephants are champions in hearing, smell and touch.

Elephants . . .

◎ Have **huge ears** mainly to help them keep cool. But they're good for hearing. They act like big funnels directing sound into the inner ear. They also help them pinpoint where a sound is coming from.

◎ Can hear very low **infrasonic** sounds and can call to each other over long distances. They can hear and recognize the calls of different elephants.

◎ Are contenders for the animal with the world's best sense of **smell**. That's not surprising, given the size of that enormous nose!

◎ Have their **nostrils** at the tip of the trunk and also have a second **smell organ** in the roof of their mouth, mainly for picking up smelly messages sent out by other elephants.

◎ Have a super-sensitive sense of touch. The **trunk** is

one of the most sensitive body parts in the animal kingdom. It can feel the slightest movement or change in pressure in the air and very carefully explore anything it touches.

◎ May even be able to feel an **earthquake** before it happens.

As well as using their trunks to touch and smell, elephants use them for all sorts of tasks – lifting, grasping, eating, drinking, breathing, sucking objects and trumpeting. The trunk has 40,000 muscles.

Other senses

Did you know that there are more than five senses? We have some of them, while others are special to certain animals . . .

Electrosense

Some animals can sense the mild electrical pulses that are given off by their prey's movements, even their heartbeat. Sharks and the duck-billed platypus can do this. Bumblebees use their electrosense to find flowers!

Pain

We feel pain. In fact, we are pretty sure that all animals that have brains can feel pain.

Earth's magnetic field

Some creatures can even sense the Earth's magnetic field. Honeybees can do this, as can many migrating animals such as sea turtles, salmon and many birds.

Balance

This sense allows us and other animals to move around smoothly without falling over. Our balance-sensing organs are in our inner ear.

Infrared

Some snakes can see infrared, which is a kind of energy given off by living things. We can feel this energy as warmth, but snakes can actually see it, like a glow lighting up their prey! The snakes have holes called pits in their head, which they use for this sense.

Body awareness

This is the sense of knowing where your body is. The scientific name for this is **proprioception**. If you do a handstand, you understand that you are upside down and your body knows what movements you need to get upright again. If you're running around, you have a sense of how close you can get to things without crashing into them, and this isn't just from what you can see and hear but from your awareness of where your body is right now.

For example, a stag is always aware of where its antlers are. This isn't like a cat's whiskers, as the antlers may be even wider than the stag's body and every year they grow a slightly different size. But they know where their antlers are. It's like me and my hair. People are always asking me if I step on it. But I know where it is! I feel when it's grown longer. I'm using my body-awareness sense for this. Dancers use this sense especially well.

Animal sense champions

 Sight

I'm awarding first prize to the **mantis shrimp**. It has compound eyes, which it can move independently, and it has sense cells for 12 to 16 different colours (compared to a human's three) including ultraviolet, infrared and polarized light.

Hearing

First prize – even better than an elephant – is the **greater wax moth**. It can even hear bats' calls, which are meant to be too high-pitched for most animals to hear, which means it can escape the bats.

Smell

My winner here is the **elephant**. Elephants can even tell by smell whether one pile of food is bigger than another – just like we would do by sight.

🥇 Taste

First prize goes to the **catfish**, with as many as 250,000 taste buds all over its body. Catfish eat pretty much anything, so they are finding food through taste, not using their taste to be picky.

🥇 Touch

I award first prize to the **star-nosed mole**, but **spiders** come pretty close.

🥇 Super-Sense

Joint first goes to the electrosensitive **duck-billed platypus**, the **pit viper** that can see infrared and the **robin**, a champion of magnetic sense.

Hamza's Top Nature People

Dr Gemma Clucas

Dr Gemma Clucas is a penguinologist – an ornithologist who spends most of her time researching penguins. I asked her a few questions . . .

Hamza: What does your job involve?
Gemma: I study seabirds, including penguins, puffins and terns. I look at what they're feeding on and how that's changing. I collect seabird poo and analyse the DNA of the creatures in it, to figure out what they've been eating.

Hamza: What is one of the best places you've travelled to?
Gemma: Probably Antarctica, because it's just such a special place – thousands of penguins, seals and whales, and lots of snow and ice, glaciers and mountains.

Hamza: What would you say to a young kid who wants to do the same job as you?
Gemma: I would say, get outside. The UK has fantastic seabirds – we've got puffins, gannets, terns, cormorants and guillemots breeding here. Later, when you choose what to study, biology, maths and computer science are all helpful.

Hamza: What's your favourite animal?

Gemma: It's got to be albatrosses. The wandering albatross has the biggest wingspan of all the birds, and albatrosses can travel for thousands of miles around the world, and they always come back to the same partner to breed, which I think is lovely.

Hamza: What's your favourite penguin?

Gemma: Adélie penguins, because they have a lot of character. Even though they're quite small, if you go up to an Adélie it will stand up tall and growl at you – they're pretty feisty birds.

Hamza: How do you cope with the smell of penguin poo? Do you wear a mask?

Gemma: No! You just get used to it. I actually love the smell of seabird poo now, because it reminds me of being in the best places around the world.

Hamza: What is the remotest place you've ever worked?

Gemma: The South Sandwich Islands. When we were there, we figured out that the next closest people to us might have been on the International Space Station.

Hamza: Do you think there is hope for our planet? Do you think we can still save it?

Gemma: Yes. I think we've seen that wildlife can be very resilient. If we give it the space it needs, it will come back.

Chapter 4
Moving

Hamza's Top Ten favourite animals

7

Back to my top ten animal countdown – and coming in at number seven, it's an animal I lived with for a little while – **the otter.** The Eurasian otter to be exact.

Thanks for looking after me, Hamza!

Where I live, everyone knows me as the animal guy, and sometimes this has put me in very unusual situations, such as when an otter came to stay!

One day, a neighbour, Julie, appeared with a baby otter in the back of her van. The first thing I asked her was, 'Do you have all your ten fingers?!' Because otters can really bite! Yes, she did, and in the van was this little tiny cute thing. He let me cuddle him. What was going on? It turned out that Julie's dog had found the cub on the beach. Julie waited a long time but no mother otter appeared.

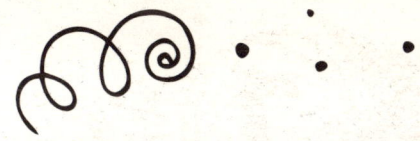

I've got a friend who has an otter sanctuary. I rang her, but she was away for a while so she told me I would have to look after the otter until she got back. This is what you do, she said: You run it a bath every day, and it will poo and pee in the bath. Keep it in a big cardboard box, putting loads of towels at the bottom for grip. So I did this, and gave it fish and water every day.

We named the otter Uisge (uuzh-gah), which is Gàidhlig (Scots Gaelic) for water, as in water of life. (I wanted to call him Midge after the otter in *Ring of Bright Water*, but my dog was already called Midge.)

I was holding him all day. He was super sweet – he just wanted cuddles all the time. At the end of the day, we get to bedtime and he starts squeaking. I pick him up and he falls asleep. I put him back down, he squeaks. Three days of this!

By the third day I'm really tired. I think I'll just lie down in bed and I'll let him fall asleep on me, and then I'll put him to bed. I fell asleep . . . in the morning I woke up and thought, did I just dream I had an otter in my bed? No! There he was! Curled up next to me looking all cute under the duvet.

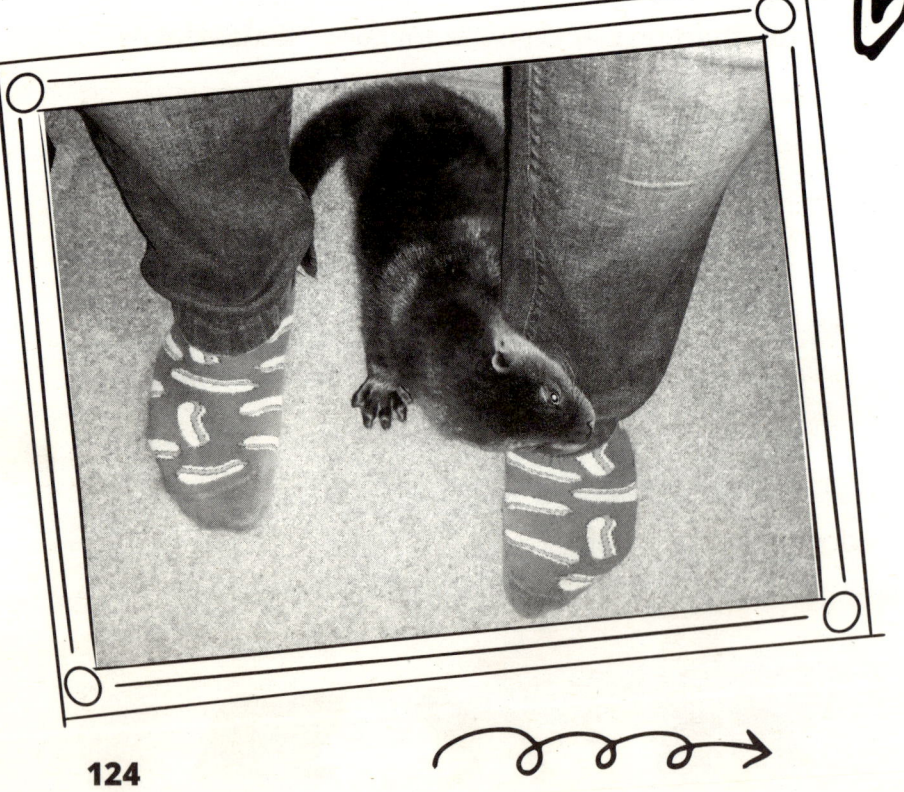

Of course, I'm supposed to be rehabilitating this otter, not taming it. So this is not what is supposed to happen! After

three weeks my friend is back from her trip and we take the otter to the vet. Immediately Uisge bites the vet! I think 'Yes!' because it shows he's still wild.

This was an unusual case. The mother had kicked baby Uisge out because he had an infected foot and the bad smell from the infection would have attracted predators to the holt, putting her and the rest of the babies in danger. So he had to go, the poor thing.

Otters have two dens. One is a maternal or nursery den called a holt, where they give birth and raise their babies, maybe a few miles up in the hills. Normally this holt will be next to a stream, under a tree or a log, where the scent is taken away by the water. But we think the mum brought this baby down to her daytime or diurnal holt near the seaside, and left it so that her other babies would be safe.

Facts

Scientific name:	*Lutra lutra*
Mammal family:	mustelids (including badgers, otters, martens, stoats and weasels)
Length:	up to 130 cm
Found in:	rivers, lakes, at the coast
Eats:	mainly fish and crabs
Babies:	2–3 (called kits or cubs) per litter
My three words:	intelligent, swimmer, cute

An otter has stiff whiskers which it uses to feel its way in the dark or in muddy water.

An otter is quick and nimble. Its long, strong tail helps push it forward and steers it as it swims. It is slim and streamlined, with a bendy body that allows it to twist and turn as it chases fish, and it has webbed feet to help it push along.

Otters can stay underwater for up to four minutes.

They need rivers where the water is clean and there is plenty of undergrowth to hide in.

Otter poo, called **spraint**, is full of information. Other otters can tell by the smell whether the animal that left it is a mum with babies, for example.

What to do if you find an animal

Uisge isn't the only unusual house guest I've had. I took care of a kittiwake that was exhausted by a storm, nursed a starling that I had accidentally singed in the chimney and trained the pine martens living in my attic to poo in a box like a litter tray.

But what should **you** do if you find a wild animal? Almost every time, the rule is to keep away. The most common example is if you find a baby bird on the ground, maybe cheeping and seeming distressed. **Leave it alone!** Many kinds of young birds find themselves on the ground before they are quite able to fly. Most likely the parent will be nearby and will come back soon to feed it and help it find a safe place to hide. Young tawny owls even climb back up into their nest – climbing down and climbing back up is a natural part of their growing up.

If the bird looks injured or has been there for a very long time without the parent appearing, or it is a very young bird with no proper feathers, the best

thing to do is call the RSPCA or SSPCA for advice. Do not feed it or touch it unless the wildlife expert tells you to. If there are predators such as cats nearby, you can try and scare them away.

Some babies, like deer, hares and seals, are supposed to be left on their own and know to lie very still to avoid attracting predators' attention. Deer are well camouflaged, too. Others, such as squirrels and badgers, should not be out of their dreys or setts and will need help. Because some animals may be injured, it's still best to keep out of the way and call the RSPCA, SSPCA or the special helpline for marine mammal rescue, BDMLR. (Or, in the case of Julie, take it to Hamza! I'm still shocked that she didn't get a nasty bite, though.)

Find out more: RSPCA, SSPCA, BDMLR, page 400

Running

Lots of animals need to be good at running – to chase prey and to escape predators. Their bodies are fine-tuned for running performance.

Humans are built for long-distance running. You may not think so when you're running around in the rain at school, but it's true. Millions of years ago, the first humans caught their prey by **running for hours**. We may not be able to sprint as fast as a gazelle or a deer, but they will get tired before we do, believe it or not!

We're efficient at running, and don't waste energy. It starts with our **feet**, with the way our ankles are shaped. We run on the balls of our feet. We have leg and foot tendons that act like springs and our feet and toes are shaped to push against the ground and propel us forward.

Our feet are very sensitive and are amazing shock absorbers. In Sudan, I have to wear *shib-shib* (which is what we call flip-flops) but people who live there go **barefoot** everywhere. They have thicker skin on their feet and their feet hug whatever surface they step on so that they can grip it properly – even in the slippery Nile – and don't hurt their feet on stones. Their feet are strong and flexible and they can even grip when climbing, by splaying out the toes. When we wear shoes all the time we miss out on developing this amazing quality in our feet.

It's not just feet that make us good runners. We are upright so we can see a long way, and we have hairless faces and bodies that lose heat and stop us overheating when we run. Lots of features help us keep our balance, including our **big bottom** that keeps us stable when we move our legs.

So, we may not be the fastest, but humans are very efficient animal runners.

Flight

Flight is one of the most incredible animal skills. Humans admire it so much that we have spent huge amounts of energy inventing machines to allow us to do it. Three groups of animals have evolved to be flying machines: insects, bats and birds.

Insects are the only invertebrates that fly (although not all of them can). Flight is their superpower. Some insects can fly vast distances. Some can beat their wings at incredible speed (a honeybee beats its wings nearly 200 times a second).

Hoverflies and **dragonflies** can hover in mid-air, using their wings to make a figure-of-eight, a bit like a rower sculling to stay on the spot. Dragonflies can also hover and fly backwards. Butterflies fly slowly and in a wobbly, unpredictable manner, probably to make them harder for predators to catch.

Some insects have four wings and some have two. **Houseflies** have two real wings and two thin stubby structures called halteres, which help them balance and keep a steady wingbeat as they fly.

Bats are the only mammals that can truly fly, not just jump or glide or soar. Bird wings are like our arms, but bat wings are like giant hands with really long fingers. Like our hands, they are very flexible and the bat can move each finger individually. This makes bats great at performing twists and turns in flight – even better than birds. Their wing bones are thin and light compared with other mammals' arm or front leg bones, but they are strong. The wing itself is just two thin layers of skin full of blood vessels and nerves.

Flying insects and birds generally have very good eyesight. Bats, on the other hand, use hearing to help them fly, and they have some of the best hearing in the animal kingdom.

Find out more: Animal eyes, page 90

Spotlight on

Dragonfly – expert flier

Dragonflies are some of the most expert fliers in the insect world. **Dragonflies . . .**

- Have **four** wings that can move independently from each other, making them extremely agile.

- Can fly **backwards, hover** on the spot or even fly **upside down.**

- **Catch** their prey in mid-air and eat it on the wing.

- **Dance** in the air to attract a mate.

- Have **huge compound eyes** that allow them to seize their prey with deadly accuracy.

- Are **fierce** hunters. They are successful, too – they catch what they are chasing nine out of ten times!

- Are some of the planet's **oldest** fliers. They evolved around 300 million years ago – even before the dinosaurs were alive.

The movement of dragonfly wings helped people design helicopters, which can also hover, fly backwards and change direction quickly.

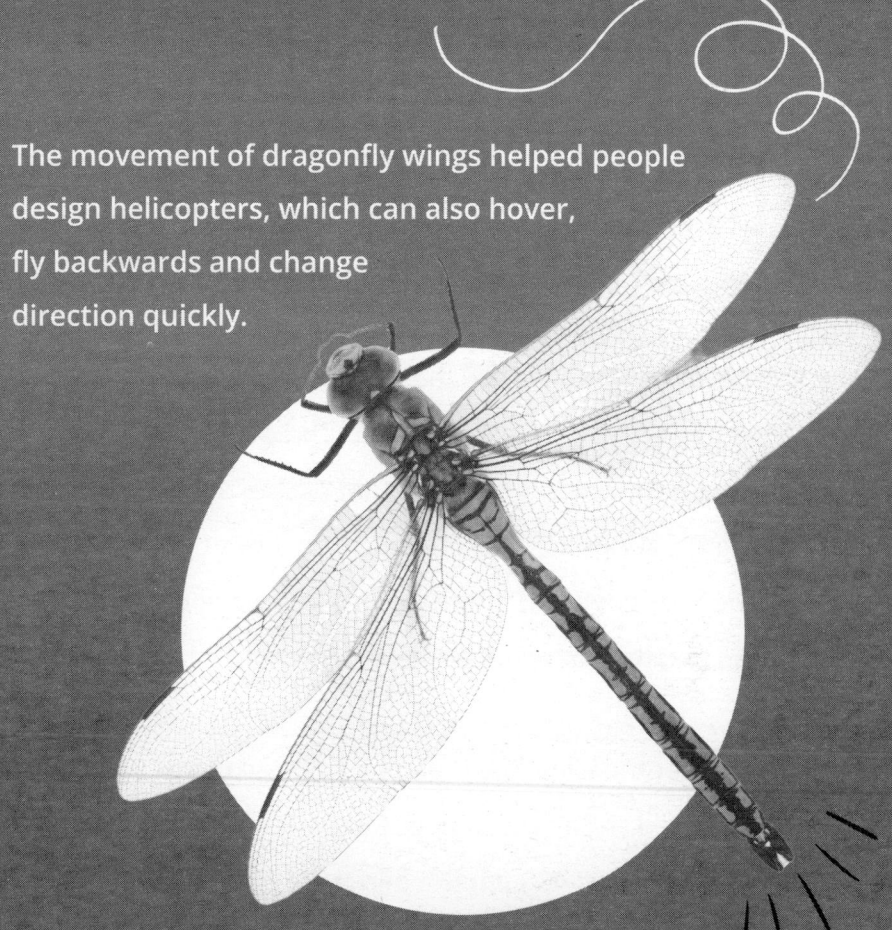

In some places, thousands of dragonflies roost (or sleep) in one area. When I was filming *Strictly Birds of Prey* I joined fellow wildlife cameraman Simon King on the Somerset Levels at dawn to film thousands of four-spotted chasers waking up. It was a truly stunning sight. The wetland was filled with clattering wings. Only the fastest, most agile predators can catch dragonflies in flight. The hobby (one of the smallest birds of prey here in the UK) is one of the only ones that can do so.

Birds in flight

We've already mentioned that bird bodies are built for flight. Powerful wings do most of the job. The wings are shaped to create lift, which is the force that pushes something upwards. Wings are thicker at the front edge than they are at the back edge and they have a pointed shape. The fastest flyers and the fastest divers have narrow, pointed wings.

The larger the wing, the greater the lift – so birds with small wings need to flap their wings faster. Puffins, for example, beat their wings 400 times a minute to stay in the air.

Many birds glide for long stretches. Instead of beating their wings, they simply hold them outstretched at an angle that allows them to stay in the air and keep them moving forward. Albatrosses can glide for hours. They use less energy gliding than sitting down!

Where there is a current of air moving upwards, called a thermal, some birds can ride these currents without having to use lots of energy flapping. You might see buzzards doing this. Vultures in Africa are famous for doing it, too.

Birds' wings are covered in **feathers**. Feathers are made of keratin, a tough, light material (the same as our fingernails and hair). There are different feather types, including soft, fluffy feathers called down, flat body feathers, tail feathers and strong, straight flight feathers.

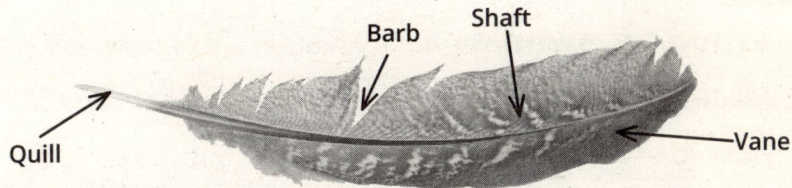

The flight feather is wider on one side than the other, which gives it an aerodynamic shape. The barbs (individual strands), which act like Velcro, clinging together making a smooth, unbroken surface to push against the air. If you stroke a feather from quill to tip, you'll feel it is smooth. If you stroke it the other way, tiny hooks on the barbs detach and you'll rough up the smooth edge.

Woodcock feather – a really beautiful feather. The brown pattern is lovely and I can age the particular woodcock just from its feathers, as they have little markings that develop a buff tip at the edge as they get older.

Whooper swan secondary feather – this pure white feather is from the second layer of flight feathers on the outside of the wing. These are fantastic, but you can see in this feather that the bottom is thicker than the top – this is because it is wearing out from flight. When the feather was new, it was nice and round all the way along, but over time it narrows so much that eventually there is no lift. When feathers are worn out, the swan drops them and grows new ones.

Find out more: Parts of a bird, page 36

White-tailed eagle secondary feather – even bigger than the swan's! It's also stiffer, because it's a bigger, much more powerful bird. It can't have a floppy wing. Think of road cyclists: their bikes never have suspension – they have a rigid bike, because with every push they want their effort to go straight to the back wheel; they don't want the effort to be wasted by being spread out through the suspension. It's the same with the eagle feather.

Barn owl flight feathers – the other extreme. Barn owls hunt by ambushing their prey. The design of this soft, bendy feather allows the owl to swoop down onto its prey in almost total silence. If swans fly over your head, you will hear their wings make this 'whoop whoop whoop whoop' noise. But an owl doesn't want the mouse to know it's coming, so the edge of the feathers are serrated, with little cuts and divots in them, which means it can swoop through the air without making a sound. Humans have copied this idea to make wind turbines quieter.

Birds of prey in flight ...

Golden eagle
Wingspan 190–230 cm

Red kite
Wingspan 175–195 cm

Andean condor
Wingspan 300–320 cm

Harpy eagle
Wingspan 176–200 cm

Peregrine falcon
Wingspan 95–115 cm

White-tailed eagle
Wingspan 180–240 cm

Osprey
Wingspan 150–180 cm

White-backed vulture
Wingspan 218–220 cm

Buzzard
Wingspan 113–128 cm

Sparrowhawk
Wingspan 55–70 cm

Ospreys and red kites are UK reintroduction success stories. The osprey is an expert fish hunter whereas the red kite is a successful scavenger.

The Andean condor is the world's largest bird of prey. It's a vulture found in the Andean mountains of South America.

When they dive, peregrines have a teardrop shape. We as humans have copied this shape for some very fast bomber planes.

141

The return of the white-tailed eagle

I am lucky enough to see the magnificent white-tailed eagle – one of Europe's largest birds – near my home. I owe this fantastic experience to the hard work of conservationists, and one man in particular.

White-tailed eagles went extinct in Scotland in the early twentieth century, and in England long before that, because people illegally hunted them.

It's hard to understand why people would want to harm these creatures but farmers feared they would kill lambs and gamekeepers believed they took grouse. In fact, although they eat live rabbits, geese, fish, seabirds and the occasional dead lamb, a lot of the time these eagles don't even hunt. They scavenge, clearing up dead carcasses and taking the leftovers of other predators' kills.

In 1975, another of my heroes, conservationist Roy Dennis, introduced several eaglets, brought over from Norway, to the Isle of Rum in Scotland. In 1985 a pair successfully bred on another island, the Isle of Mull, for the first time. Since then, the birds have established themselves and in 2023 there were around 150 breeding pairs.

Conservationists have given the eagles a helping hand. The white-tailed eagles usually have two or three chicks at a time, but often only the strongest chick will survive. Taking one chick and rearing it in captivity ensures that the chick is cared for and gives the remaining chicks a better chance to find enough food.

In ancient times eagles soared across the whole of Britain – white-tailed eagles in the coastal areas and golden eagles in the mountains. They were a missing piece of the wildlife puzzle in the UK and it's great to have them back.

Swimming

Let's meet some animals that are great movers in water.

Gentoo penguin

Lives: on pebbly shores in the south of South America and in Antarctica

Body: extra-streamlined, even compared with most birds. Wings are more like flippers. Feathers are very short and close together, making the birds almost waterproof

Swims: by using its flippers as rudders and propellors to steer and power it through the water

Amazing fact: this is the fastest penguin and it can swim underwater at 35 km/h!

Gannet

Lives: on coasts of the northern Atlantic Ocean during the breeding season and out at sea for the rest of the year

Body: closes its nostrils when diving so it doesn't get seawater rushing up its nose. Also has a strong neck and built-in airbags so it doesn't get concussion with the force of hitting the water

Swims: uses its powerful wings and impressive speed to dive at 88 km/h

Amazing fact: it can catch fish 20 metres below the surface!

Grey seal

Lives: on coasts of the northern Atlantic Ocean during the breeding season and out at sea at other times

Body: torpedo-shaped body. Legs are short, stocky flippers and body has a thick layer of fat to keep warm underwater

Swims: uses its back flippers to propel its body along and its front flippers as rudders

Amazing fact: has super-sensitive whiskers to help find fish in murky water

Green turtle

Lives: oceans worldwide except for polar regions

Body: front legs are long, paddle-like flippers. Can stay underwater for up to five hours because it can slow its heart rate right down to conserve oxygen

Swims: uses its front flippers to propel its body along and its short back flippers as rudders

Amazing fact: females return to the same beach year after year to lay their eggs, after swimming thousands of kilometres throughout the year!

Atlantic herring

Lives: throughout the Atlantic Ocean

Body: streamlined, fish do not have arms or legs

Swims: powers forward by moving in an S-shaped sideways motion and uses fins to steer and change direction

Amazing fact: this is the animal champion of synchronized swimming . . . hundreds of millions of them swim together in perfect harmony!

Common octopus

Lives: in seas from the south of England to the north of Africa

Body: soft, with a cavity filled with water

Swims: uses jet propulsion, squirting jets of water from its body at high speed and pushing itself backwards through the water

Amazing fact: can jet along at up to 40 km/h!

10 Fast Animals

The pronghorn antelope is the fastest long-distance runner. It can run at a steady speed of 56 km/h for 6 kilometres, whereas a cheetah can run up to 105 km/h but only for about 200 metres!

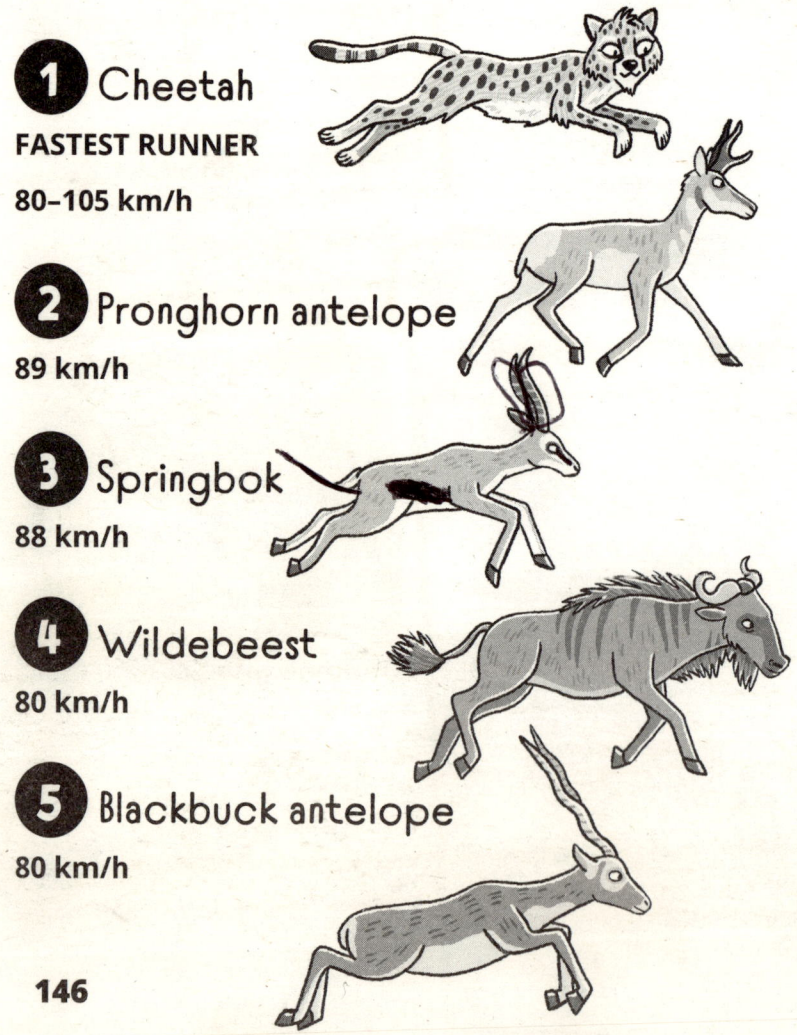

1 Cheetah

FASTEST RUNNER

80–105 km/h

2 Pronghorn antelope

89 km/h

3 Springbok

88 km/h

4 Wildebeest

80 km/h

5 Blackbuck antelope

80 km/h

6 Lion
74 km/h

7 Brown hare
72 km/h

8 Thoroughbred horse
71 km/h

9 Thomson's gazelle
64 km/h

10 Elk
64 km/h

Fastest dive	**Fastest straight flight**	**Fastest swimmer**
Peregrine falcon	Mexican free-tailed bat	Sailfish
320 km/h	**160 km/h**	**109 km/h**

147

Jumping and climbing

Jumping takes a lot of energy, so why do kangaroos prefer to get around in leaps and bounds? The answer lies in their tendons (taut cords that link muscles to bones). They use their tendons in their leg like a spring, to store and release energy very efficiently. Kangaroos also use their strong, bendy tail to balance and push themselves into the hop. It's very energy efficient.

Some animals jump so well that they seem to fly. The flying squirrel, flying frog, flying dragon lizard and colugo all jump and **glide** from tree to tree or branch to branch in a movement that looks a lot like flying. Usually they have a flap of skin linking some of their limbs or digits (fingers and toes), making something that looks a bit like a bat's wing. They can't lift themselves off the ground or beat their 'wings' but they can glide and sometimes steer. A flying squirrel can glide for nearly 100 metres and turn in mid-air.

Climbing is a complicated movement, but many animals are masters of this skill. Apes (including us) and monkeys are a good example. We have shoulders that can move in different directions, flexible elbows, wrists and hips, and very dextrous, gripping hands and feet. We can swing (think of the monkey bars on the climbing frame) and we are intelligent. Monkeys also use their tails as an extra limb both for balancing and to wrap and grip.

Geckos have amazing sticky pads on their feet which allow them to cling on to and climb up any surface easily.

Red squirrels have incredible brain capacity, which allows them to assess the position and the strength of a branch in milliseconds, so that they can climb right to the end of a thin pine branch or leap from tree to tree.

Stealth

Whether they are moving as fast as they can, walking slowly or even staying still, many animals are experts in stealth movements.

Have you ever noticed how **cats** walk? They step with a front paw, then carefully place a hind paw into exactly the same footprint they trod on with the front paw. This muffles the sound and helps them walk silently. If, by mistake, their front paw hits something that makes a noise, they will immediately **adjust** their step so that the hind paw doesn't land in the same print.

Barn owls have specially engineered feathers so they can fly silently and pounce before their prey knows they're even there.

Crocodiles and **alligators** can stay motionless on the surface of the water, making it hard for their prey to notice them. They can also dive, roll and swim so

Find out more: Barn owl feathers, page 139

smoothly that they hardly make a ripple. When they do pounce, they do so quickly and with a **splash!**

Eagles use the Sun as a stealth aid. They often position themselves in a way that means prey will be dazzled by bright sunlight, so that they are unable to see the eagle approaching. The wedge-tailed eagle from Australia is the master of this stealth-and-distraction technique, and they work in pairs. One distracts the prey, and the other swoops in while its prey is blinded by the Sun.

Coyotes walk on tiptoe, like a person might if they were trying to walk very quietly.

Herons are masters of stealth. Have you seen one standing perfectly still, watching for prey? They can do this for hours. Some species of heron even use their wings to make a shady canopy and fool creatures into thinking it is night-time.

All these animals use camouflage for stealth, too – but stealthy movement is vital to their success.

Animal tracks

As animals move through the landscape, they leave clues behind. Footprints, for example, tell us what animal has been there and give us quite a lot of information about how they move. You can do some detective work by looking out for these tracks and other clues.

Tracks

Tracks are footprints and other marks that animals leave behind on the ground. Look for them where the ground is muddy or after snow. Look for features such as how many **toes** there are, and whether you can see **claws** or **pads** (which dogs have).

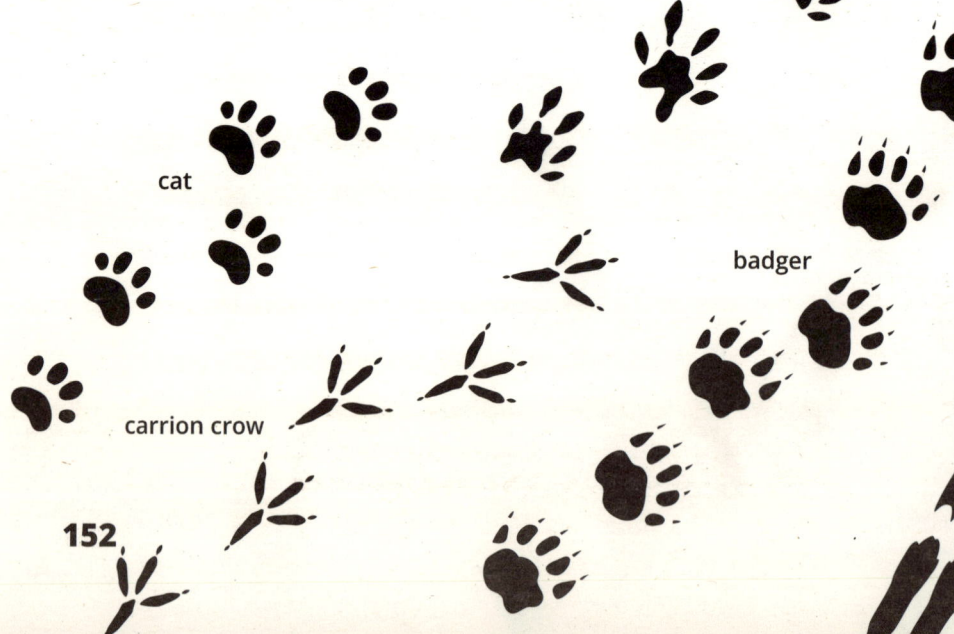

cat

badger

carrion crow

fox

heron

otter

duck

dog

rabbit

Other clues

Footprints are not the only clues.
Look out for tufts of fur (especially
on barbed wire or thorns), feathers,
broken eggshells or remains of meals,
such as nibbled nuts or gnawed bark.

For clues about poo: see page 234.

Migration

One of the most amazing kinds of animal movement is migration. Animals on the move make journeys of hundreds or even thousands of kilometres. Lots of people think migration is just about birds, but it's so much more than that. Tuna, sharks, rays, crabs, butterflies – they're all on the move. Here are some of my favourites:

- The red-necked phalarope travels from the north of Scotland to North America near the Arctic Circle, then all the way down to South America. It's a very unusual bird because only the male sits on the eggs. The females lay a few eggs in different males' nests, then set off on their migration.

- Swifts spend the winter in central and southern Africa then migrate to their breeding grounds in northern

Europe, including the UK. You'll see them speeding high in the sky and hear them screeching from May to August. They're supreme flyers: they eat, sleep and mate on the wing, and only land to nest. They can fly as high as 4,000 metres in some parts of the world and can travel 800 kilometres in a single day. They can even fly through storms without getting blown off-course.

One of the most amazing migrations is that of the wildebeest in the Serengeti, in Africa. I've filmed it and it's mind-blowing. For days on end, thousands and thousands of wildebeest keep coming, walking through the plains from areas of short grass to areas of long grass, following the grass and the rain. More than 1.5 million wildebeest go on this journey, walking in an enormous never-ending loop through parts of Kenya and Tanzania.

I would love to film the kob migration in South Sudan. Kob are small antelopes that migrate in herds of over a million individuals from place to place on the Nile plains of South Sudan. This migration is as big as the wildebeest migration but hardly anyone knows about it – partly because the civil war there has made it difficult to study them.

Eel migration has long been a mystery. We know that European eels arrive in European waters as tiny, young glass eels. When they grow up they migrate across the Atlantic Ocean to part of the Caribbean called the Sargasso Sea, where they spawn and lay eggs. Using satellite tags, we are only just finding out how they get there and what route they take.

One of the longest animal migrations is that of the globe skimmer dragonfly. It makes a round trip of an incredible 18,000 kilometres, from India to east Africa and back, including flying across the Indian Ocean and stopping off in the Maldives and the Seychelles. It takes more than one generation to make the journey – the dragonflies that arrive back in India are the great-grandchildren of the ones that set off in the first place.

When I first moved to England I saw birds that I recognized from my home in Africa. Sights such as this bring home the idea that birds have no borders and don't just belong in one part of the world. Bad weather, habitat loss, hunting or other problems in one part of their migration route can affect wildlife populations thousands of kilometres away.

Migration
The Manx shearwater

Animals migrate to take advantage of rich food resources and good breeding conditions. For example, redwings and fieldfares migrate to the UK in winter because it is too cold further north in Scandinavia. In summer they travel back to Scandinavia to breed.

Animals use various skills to find their way. Some use the position of the moon, sun and stars; others follow landmarks, such as rivers or mountains, and build mental maps; still others follow weather or the smell of grass. Some even sense the Earth's magnetic field and get their direction from this.

One of the most remarkable migrations I have filmed is the journey of the **Manx shearwater**. These small seabirds use the stars to navigate.

Manx shearwaters nest in colonies of hundreds of thousands of birds on small offshore islands such as Rum in Scotland and Skomer, Skokholm and Middleholm

in South Wales. In July, just before they're ready to set off on their long migration, the young birds come out at night, sit on the ground, staring up at the sky and making an eerie wailing noise. People once thought those islands were haunted.

During this strange night-time gathering we think they are stargazing – using the stars to work out exactly where in the world they are and to store up a mental map for later, when they'll return to this place year after year. Or perhaps they find their way using the Sun's position and the Earth's magnetic field – we don't know for sure. When they're ready to fly, they set off on a mind-boggling flight all the way across the Atlantic to Brazil and Argentina (over 10,000 kilometres), where they spend the winter. Some of them take as little as a fortnight to get there.

Hamza's Top Camera People

Jesse Wilkinson

Jesse Wilkinson is a fantastic wildlife cameraman whose work includes *Our Planet II* and *Wild Isles*. Working for him was the break I needed. I asked him a few questions . . .

Hamza: How did you start off as a cameraman?

Jesse: I was born in Manchester. I remember everything being very grey and dull. The only thing that caught my attention was a little blue tit that came to the garden bird feeder. For me at five years old, it was just the most colourful thing I had ever seen. Soon after, I moved to Wales and just fell in love with nature. Like you Hamza, I'm dyslexic. But being dyslexic doesn't get in the way of watching nature. Everything makes sense if you watch carefully and I became not just very good at identifying animals and birds but being able to anticipate what they were going to do next. After leaving school I worked in conservation, but it wasn't quite right for me. I decided to put my wildlife and artistic skills into filming. I got extremely lucky and met Simon King, the

cameraman and presenter, and he was kind enough to give me a job on *Springwatch*. At last, I had found something I loved doing and I was really good at!

Hamza: What was it like being able to capture the sequence of the white-tailed eagles hunting geese?

Jesse: I know Islay well and I had seen the eagles hunting geese. It's an extraordinary thing to witness. I was able to convince the producers of *Wild Isles* to try to film it. The first couple of years were quite frustrating but everything came together in the third year. It was amazing. We erected tower hides and the eagles hunted right in front of us!

Hamza: What would you say to a youngster who wants to become a camera person?

Jesse: You can learn about cameras . . . but really understanding animals is key to getting good shots.

Hamza: What is your favourite place to film?

Jesse: Svalbard, in the Norwegian Arctic. It puts you in your place. It makes you feel very small and insignificant, and the wildlife is just unbelievable.

Hamza: What's your favourite animal to film?

Jesse: I love filming polar bears but any animal doing something interesting and engaging or in beautiful light is a thrill.

Hamza: What is one thing you would do to improve the future for wildlife?

Jesse: If I were prime minister, I would make sure our cities were greened. Every building needs plants and trees growing on them. Wildlife would come back. I think people would start to be happier and would care more about the world around them. Singapore is a good example. But everyone can do something even if it is just planting flowers in a pot on your window sill for insects to come and feed.

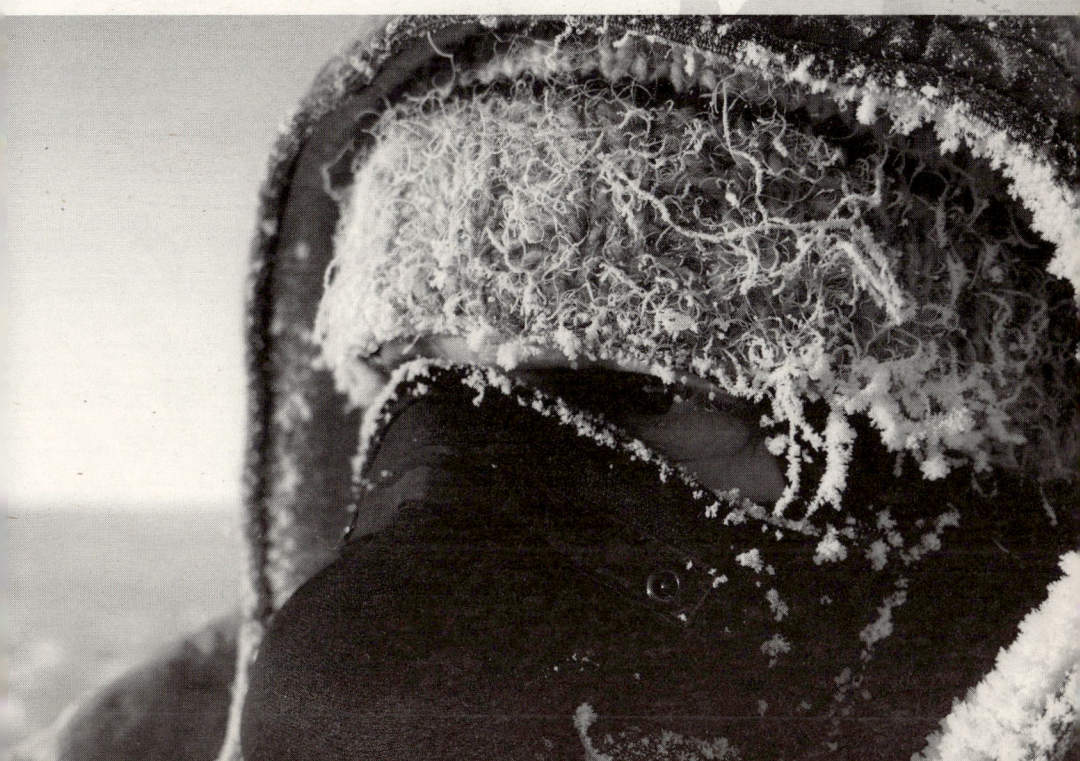

Chapter 5

Fighting and Getting Along

Coming in at number six is a wonderful ocean mammal, the **orca**.

Orcas are social animals. They live together in family groups called pods or matrilines. Orcas often stay in their pod for their whole lives. The pod consists of a few closely related families, each with a mother in charge.

Orcas are the biggest kind of dolphin. Like other dolphins, they hunt using echolocation. They work together to hunt, using some impressive techniques.

One of them is stealth. The *Wild Isles* team filmed this in the Shetland Islands, where the orcas have an interesting way of hunting seals. The pod will get together and approach the seals and make an attempt at a hunt. The seals escape and hide in the kelp – they think that the danger is over. Normally an orca swimming along will break the surface with its dorsal fin sticking up, but now

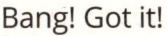

the orca comes in sideways towards the shore, getting as close as it can to the seals in the kelp without being seen. Then suddenly it appears – and the seal is in trouble! Bang! Got it!

Another hunting technique, in North America, is for the orcas to vocalize loudly as they swim away, as if they're shouting 'I'm off, bye', making any seal think that the coast is clear. But one orca is left behind, hiding. It goes completely silent, breathing really slowly, moving stealthily and – bang!

In Antarctica, the orcas do something called bow-wave washing. They all come together as one and at the last minute they dip under a raft of sea ice with a seal on it. They create a bow wave that hits the ice and washes over the top, knocking the seal off the ice and into the sea where they can catch it.

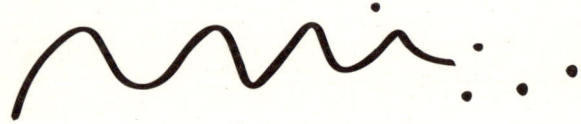

In New Zealand one particular group of orcas has learned to use their tails to stun stingrays. Stingrays can cause a lot of pain to orcas, but these orcas use their tails to shoot a large volume of water at the ray and flip it over. If you flip a ray over it goes into an unresponsive state. At this point the orca can pull the stinging barb off and then it's free to eat the stingray without getting hurt.

We can tell individual orcas apart. The white-and-black pattern on their face and near their tail (called the saddle patch) is different for every animal, and the dorsal fins, which stick up out of the water and are often the main thing you can see when you spot orcas, have slightly different shapes, with distinctive notches, nicks and scars. The marine biologists who study orcas can recognize and name each individual orca and follow them throughout their life.

Orcas can recognize each other, too. Each has its own distinctive call which is like a signature or name.

Here in the west coast of Scotland we have two famous orcas, called John Coe and Aquarius. These two bull orcas travel up and down the west of the British Isles. Every now and then you see them go over to Aberdeenshire, but the west coast is their territory. It's beautiful that we can get to know the individual animals. John Coe is really famous. His fin has a notch like a bite mark out of the bottom. A friend of mine even has a tattoo of John Coe's dorsal fin!

John Coe and Aquarius come from a family of orcas where there have been no calves for many years and whose females are now infertile. This is because of pollution building up in all the ocean creatures that the orcas eat. Sadly, this means that this family will die out and then I think the Norwegian orcas will start to come south to Scotland and take the vacant property and make this their territory.

Facts

Scientific name:	*Orcinus orca*
Mammal family:	dolphins
Length:	up to 9.8 m
Found in:	oceans and seas all over the world
Eats:	fish, squid, seals, seabirds, even whales. Orcas are the only animals known to eat great white sharks!
Babies:	one at a time, usually between 3 and 10 years apart
My three words:	social, clever, hunter

Orcas are successful apex predators (they do not have any natural predators of their own). We think of them as being fierce hunters of seals, porpoises and even sharks, but many of them eat mostly salmon, squid and herring. Just as some human societies eat a mainly vegetarian diet, whereas others eat mainly meat, different orca populations have different diets.

The orcas I watch in western Scotland eat mainly other marine mammals such as seals, minke whales and porpoises.

Another name for orca is killer whale. It comes from the fact that it sometimes kills whales (it's a killer of whales, not a whale that kills!).

The orca is the 'Lord of the Ocean' for the Indigenous peoples of the northwest coast of Canada, and it's sacred in Hawaiian and Māori culture.

Orcas are fast swimmers and can travel 160 kilometres in a day.

Animal groups

Living in a group gives animals safety in numbers and allows them to work as a team: hunting, keeping watch, protecting each other, sharing food and raising young. Here are some names for groups of animals . . .

a band of gorillas

a cackle of hyenas

a charm of goldfinches

a convocation of eagles

a flamboyance of flamingos

a herd of wildebeest

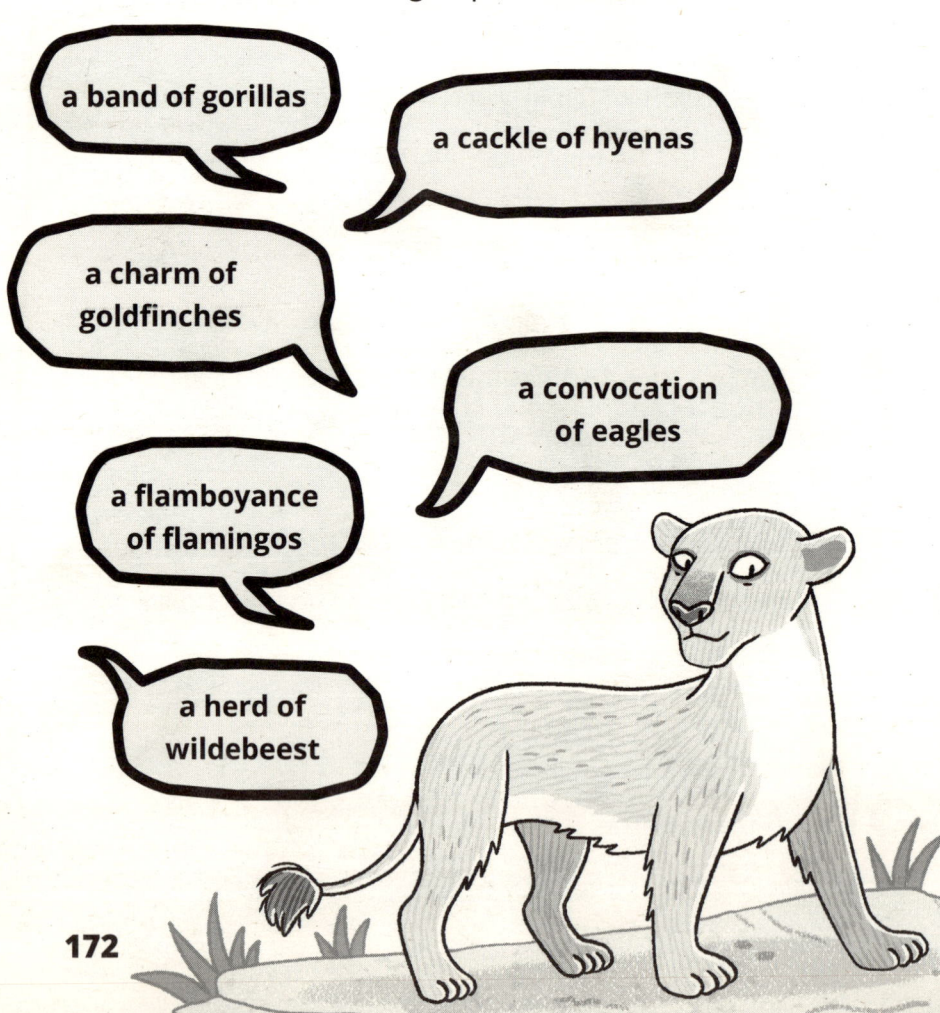

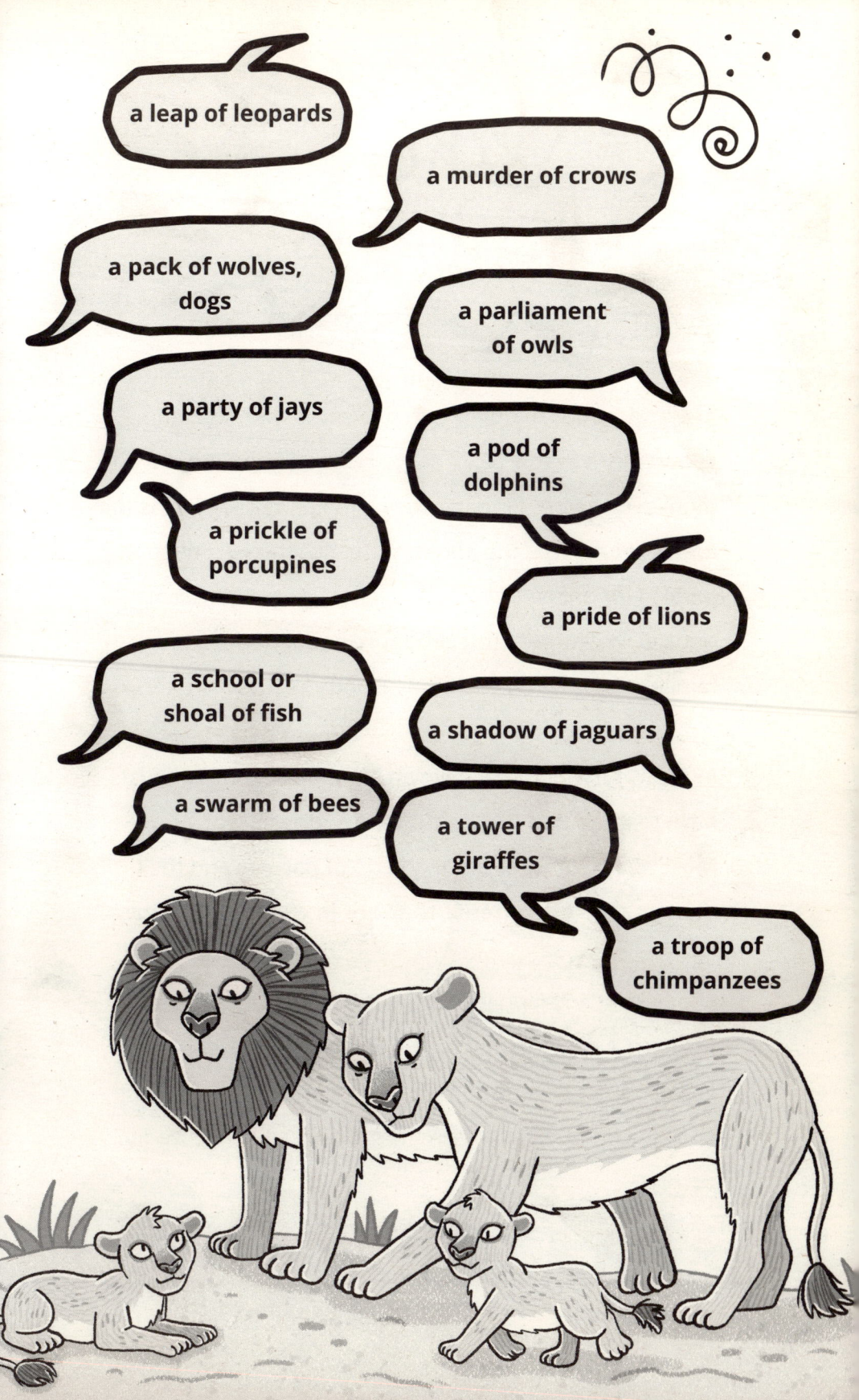

Who's the boss?

In many groups of animals that live together, the animals know who is in charge. In some kinds of apes and monkeys, the boss is a big strong male. He's known as the dominant male or the **alpha male**. But this isn't the only pattern for animals that live together. Some groups have females in charge, some are more equal, some stay in family groups all their lives, while others move around.

In captivity, wolf packs have an alpha male. But scientists have realized that this isn't the case for wild wolves. Most wild wolf packs are made up of parents and their offspring. It's the **parents that are in charge** of the pack.

The boss of an elephant herd is the oldest female in the herd, called the **matriarch**. She is the mother, grandmother or even sometimes the great-grandmother of all the other elephants in the group. It's the knowledge

she's built up in her long life that makes her the leader. Orcas are similar in that respect, with matriarchs in charge.

Insects that live in colonies – such as some kinds of ants, bees, termites and wasps – have a boss called a **queen**. She has a different body shape from the rest of the animals in the colony and they all look after her every need. She even controls them using chemical messages – stopping them from developing into queens themselves and keeping them calm and obedient.

If you have a dog in your household, it will behave according to its place in the pack order. It's a good idea that the human is 'top dog'! I don't let my dog, Midge, shake hands with people because, for a dog, placing your paw on top of another animal is showing dominance.

Who's the boss? Chimpanzees

Chimpanzees live in groups with a dominant male.

- ◎ Males compete for dominance by winning favour with the females – grooming them, wooing them – and a male knows that when he wants to become the alpha, the winner is the one the females choose to support.

- ◎ Males use babies as bargaining chips to avoid fighting. If one male gets aggressive and another picks up a baby, everyone will back off. They can't risk hurting the baby and making the females angry.

- ◎ When the male chimps get angry they puff up their fur to look bigger, start calling and also hit things with the bony lumps they have on the back of their wrists.

- ◎ Chimps are constantly building relationships, and trying to make it to the top and stay on top, but also share resources and have strong friendships.

That tickles!

- Animals of equal dominance groom each other making an A shape with their bodies, showing they are balanced and equal.
- If a chimp offends another, it places its hand in the other's mouth as an apology and to show submission.

Of the other great apes (apart from humans), gorillas live in societies with a dominant male, bonobos live in a co-operative society led by females, and orangutans are mainly solitary.

Find out more: Gorillas, page 364; orangutans, page 372

Why do animals fight?

Many animals fight among themselves. They fight to . . .

- Decide who's boss.
- Impress and get the chance to mate.
- Get the best choice of food.
- Protect their young.
- Defend their territory.
- Fight off unwanted mates.

Not getting the chance to mate is a terrible fate for an animal, because the most important thing that they can do is pass their genes on to the next generation.

Capercaillie

Size ★★	4.1 kg
When they fight	spring to midsummer
Who	several males
Why	to compete to mate with females
What happens	dancing, strutting with fanned tail, wheezing, making clicking, popping and squawking sounds
Similar to	jousting
Weapons	sharp beaks, wings
Danger rating ★★	losers don't get to mate that year

Red kangaroo

Size ★★★	up to 90 kg
When they fight	all year
Who	two or more males
Why	to be the boss and to mate with females
What happens	jabs, slaps, pushes and punches with front paws, and kicks with back feet
Similar to	kickboxing
Weapons	very powerful legs, with speed and agility
Danger rating ★★★★★	occasionally fight to the death

179

Stag beetle

Size	up to 6 g
When they fight	spring
Who	two males
Why	to compete to mate with females
What happens	antler clashes, wrestling and headlock, pushing and lifting opponent off its log
Similar to	wrestling, judo
Weapons	strong antlers (which are really their jaws)
Danger rating ★★★	disaster if it doesn't win any fights and can't pass on its genes

Elephant

Size ★★★★★	up to 6,100 kg
When they fight	when a female is ready to mate and more than one male is nearby
Who	two males
Why	to compete to mate with females
What happens	aggressive trumpeting and growling, violent clashing of heads and tusks, twisting trunks around each other
Similar to	ultimate fighting
Weapons	tusks, sheer body strength
Danger rating ★★★★★	sometimes fight to the death, also trample any animals in their way

Red-necked phalarope

Size ★	up to 40 g
When they fight	spring
Who	two or more females
Why	to compete to mate with males and guard their male from other females
What happens	smacking, dive-bombing, pushing
Similar to	scrapping
Weapons	wings
Danger rating ★★	losers may get to mate, but not with the best or healthiest male

Hare

Size ★★	up to 5 kg
When they fight	spring (think of 'mad March hares')
Who	female and male
Why	female fighting off an unwanted male
What happens	jabs and punches with front legs
Similar to	boxing
Weapons	front legs, big front paws and speed
Danger rating ★★	losers will get the chance to mate, but can't choose the best, healthiest mates

The red deer rut

One of the most intense animal fights happens on the heaths and moors near where I live in the west coast of Scotland. It's the red deer rut.

Every autumn, red deer stags (the males) square up to each other in a dramatic contest. They confront each other and show off to the hinds (the females).

The stag goes through a few changes to get ready for the rut . . .

- ◎ His body makes more of a hormone called testosterone.
- ◎ His coat gets darker from wallowing around in the mud.
- ◎ He wallows in his own wee to make himself really smelly (it's like aftershave!).
- ◎ His neck muscles get bigger and stronger.

- If he's been hanging round in a group of bachelors for the summer, he goes off on his own.

When the rut kicks off, the stags . . .

- Roar and bellow.
- Walk alongside each other in a threatening way.
- Stamp fiercely.
- Lock antlers, and shove each other and fight.

The most powerful, strong and intimidating stag will become the dominant male. He will be the only stag who gets to mate with all the hinds and pass on his genes to the next generation.

Stag antlers through the year

Deer antlers are made of bone. But unlike with most bones, deer grow new antlers every year.

Here is how red deer stags' antlers grow from one rutting season to another:

March/April:
antlers fall off

May:
new antlers, covered in soft velvet, slowly start to grow

June/July:
antlers grow about 5 cm per week – they are hardening from soft into hard bone

August:
antlers are fully grown and velvet has fallen off

The colour of antlers depends on where stags are in the landscape. Some of them are brownish, because the deer have been whacking their antlers on tree bark. In Scotland they're a deep brown, because they stick them in the muddy peat bogs.

September:
antlers are fight-ready

Hamza's Favourite Camera People

Erin Ranney

Erin Ranney is an amazing wildlife cinematographer. She is based in Alaska, Washington state and the Falkland Islands. She is a great storyteller and is especially known for her work with brown bears. I called her in Alaska . . .

Hamza: How did you start off as a camerawoman?
Erin: At first I wanted to be a vet – growing up in a small town in Alaska, I figured that was the only job you get to do with wildlife. Then at university I changed to ecology. After graduation I went to Madagascar for field research, studying lemurs. I picked up a cheap camera and completely fell in love with filming. I did a masters in wildlife filming in England, and the rest is history.

Hamza: What was your first break?
Erin: I was volunteering at a film festival and met Mark Emery, a cameraman who worked in Alaska and Florida. I showed him some of my work – he saw potential and said I could intern with him.

Hamza: What would you say to a young girl who wants to be a camera person?

Erin: I'd tell her to get outside! Learn to interact with the outdoors and animals. Are you comfortable camping? Can you pee in the woods? Know that it's OK to make mistakes. Just own up to it and move forward.

Hamza: What's the most dangerous encounter you've had with wildlife?

Erin: I would say with people has always been the worst for me! Everyone always thinks it's the bears that are the problem . . . I always say, bear spray is never for the bears.

But also, don't get cocky around bears. At the end of the day they are a huge animal that could kill you, but if you treat them with respect and read their body language, you're usually OK. The key is to not surprise a bear. The best way to do this is to make noise. That's why you sing, say 'Hey bear', or just have loud conversations with whoever's with you, just to let bears know you're there.

The monarch of the pine forests

Deep in the forests of Scotland's Cairngorms region is an incredible, majestic bird: the capercaillie. This turkey-sized member of the grouse family is found throughout northern Europe wherever there's ancient pine forests.

Capercaillies were hunted to extinction in the UK in 1785. They were reintroduced in 1837 but remain critically endangered, with only about 550 birds left in the wild.

One of the most phenomenal things to watch is the capercaillie lek, which is a gathering of up to five or six males displaying (showing off) and competing for females.

I was lucky enough, once, to film the lek. Very early one morning, in a secret location, we walk past 900-year-old oak trees and my guide brings me to a fork in the path. Capercaillies normally like a clearing for their lek, but because here there is only one

male and two females, he can't clear much of the area on his own, so he has chosen the path for his lek.

There's this weird, dinosaur-like sound echoing through the forest. It's fantastic. I get low to the ground to try and get a really good eye-level shot of this male displaying. He sees me, comes right towards me, fully displaying for about ten minutes, getting right up close, as if I were another male capercaillie! He flares his tail like a fan, sticks out his beard feathers, and makes aggressive gurgling, wheezing and popping sounds.

Conservationists have been working hard to protect capercaillies. All this effort might be starting to pay off – in 2023 Scotland had 19 more male birds than there had been the year before. But there is still a long way to go to restore these magnificent birds to their previous numbers.

Spotlight on

African buffalo – life in a herd

African buffalo know that living together is the best way for them to succeed. Buffalo . . .

- ⊚ Live in **female**-dominated herds, made up of clans – family groups of females who are related to each other, along with their offspring and a few males. Other **males** live in much smaller bachelor herds of five to ten animals or even roam **alone**.

- ⊚ Are guided by a few **dominant females** who make a joint decision about where to roam and lead the way each day. They make the decision in a kind of vote – each points her head towards the way she wants to walk and they go with the most popular choice.

- ⊚ Rely on the herd for **safety**. They arrange themselves in a formation for protection, with the dominant females in front, followed by mothers and their calves, then any sick, old or weak animals. The males make a ring around the whole herd. This pattern makes a good defence against lions.

190

- ◎ Don't **see** very well, so having hundreds of pairs of eyes working together in a herd is useful.

- ◎ Go to the **rescue** if an individual buffalo gets targeted by a predator. Together the herd chases the predator and frightens it off. A buffalo is around three times as heavy as a lion, so it's pretty frightening!

- ◎ Stay **close** to their **calves** for at least a year. After that the female calves stay in the herd and males leave when they are about four years old.

- ◎ Keep safe at **night** by staying totally **silent**, making it very hard for predators to pick out an individual to attack.

- ◎ Sometimes gather with other herds to make up an **enormous herd** of thousands of animals.

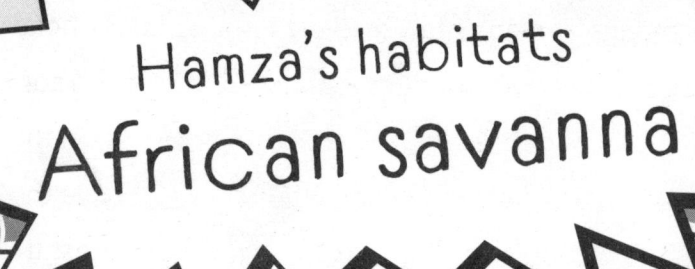

Hamza's habitats
African savanna

Habitat: African savanna

Where in the world: across huge areas of central and southern Africa (including South Sudan)

Landscape: grassland with a few trees, especially acacia and other tough, thorny trees

One of my favourite habitats is the African savanna – for example, the Serengeti. This hot grassland habitat is an incredible working machine. Just try to get your head around how many ungulates (hoofed mammals) such as zebra, wildebeest, gazelle and buffalo, walk on this landscape – it's mind-blowing. When you first see the migration, you can't believe how many wildebeest there are – thousands and thousands of them. The next day, it's still going. Days and days of wildebeest, then days of zebras. They try and hang out in the short grass, and when they've eaten that they have to move to the long grass, but that's where *simba* (the Swahili word for lion) is hiding. The sheer mass of animal life on plains and grasslands like the Serengeti is phenomenal.

I once had an amazing experience with a cheetah. I'm sitting on the truck in the savanna, watching and filming. Suddenly I hear heavy panting. There's a young cheetah next to me!

He checks out our truck – boring. Immediately he looks out over the savanna. The savanna is flat as a billiard table. He jumps up onto our truck to get a bit of height above the landscape – we're like a portable termite mound he can use as a viewing point. Once I've figured that out, I've focussed on his mum. I'm thinking, 'She's going to get up and see her baby right next to me, and she's going to come charging straight at me. I'm going to get the best shot of my life and then get eaten!' Actually, she wakes up, sees her baby is just annoying the tourists again, and goes back to sleep. So I'm OK, I can breathe again.

In the savanna, there are two seasons – wet and dry. In the wet season it is warm and rainy, and the vegetation is lush and rich. In the dry season it can be extremely hot, with hardly any rainfall, and the animals must cope with these conditions.

Did you know?

African dung beetles find their way around by looking at the stars!

Star animals of the savanna include the huge herbivores such as elephants, hippos, rhinos, giraffes, buffalo and smaller herbivores such as zebra, wildebeest and gazelles. There are impressive predators such as lions, leopards, cheetahs and some smaller cats, plus falcons and eagles. Important scavengers include vultures, which congregate in the few trees, jackals and fierce packs of hyenas. And don't forget about the African dung beetle, which deals with some of the huge piles of elephant poo!

Some of these animals live together in herds and socialize with other species, getting along for protection and to share food. Some feed on one another while others fight. But all of these animals exist together and rely on each other in an intricate network of relationships.

Animal partnerships

Sometimes, animals live together with different species. The partnerships bring benefits for both of them. This is called symbiosis.

Anemone and clownfish

This is probably the most famous animal partnership. The orange-and-white clownfish – like the one in *Finding Nemo* – lives among the stinging tentacles of the anemone.

The clownfish gets . . .

- protection from predators, which steer clear of the stinging tentacles
- protection from the tentacles (the clownfish have a layer of slime that prevents them getting stung)

The anemone gets . . .

◎ scraps of leftovers from the clownfish's meals and nutrients from clownfish poo

◎ cleaned of harmful parasites and dead bits that flake off its body (that the clownfish eats)

◎ protection from anemone-eating fish such as butterfly fish (that the clownfish chases away)

Jay and oak tree

This animal–plant partnership is an example of how intelligent some birds are. The jay collects acorns in the autumn and buries most of them. It digs up some to eat during winter but many acorns get left and grow into young saplings. In spring the jay picks some of the young, juicy green saplings and feeds them to its chicks.

The jay gets . . .

◎ rich, fatty acorns to eat throughout the winter

◎ a place to nest

◎ lots of insect larvae to eat and feed to its chicks (the larvae live in the oak tree)

The oak tree gets . . .

◎ its acorns planted and spread over a wide area – giving lots of them a chance to grow

Animal partnerships match-up

Oxpeckers
Get plenty of juicy flies, ticks and fleas to eat and a free ride on the backs of . . .

Barnacles
Get a free ride to waters rich in their plankton food by sticking to the giant backs of . . .

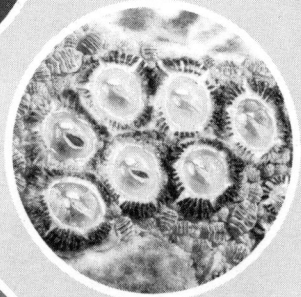

Egyptian plover
Gets leftover scraps of meat from somewhere most animals would find far too dangerous – the jaws of the . . .

Pilot fish
Gets scraps of leftover food and a fearsome protector, a . . .

Yellow ants
Get a sugary liquid called honeydew by 'milking' . . .

The yellow ants also get honeydew from the larvae of . . .

. . . Buffalo
Which are kept clean and disease-free by the small, colourful birds and rely on them to raise the alarm when a predator approaches.

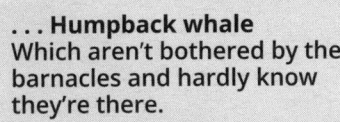

. . . Humpback whale
Which aren't bothered by the barnacles and hardly know they're there.

. . . Nile crocodile
Which manages not to snap its jaws on the brave bird, and lets it clean its teeth!

. . . Shark
Which gets its parasites gobbled up and its teeth cleaned, as well

. . . Butterflies
Which, as larvae, get protection from parasites and predators (including other ants!).

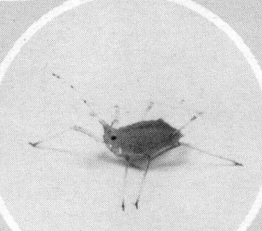

. . . Aphids
Which get looked after and protected by ants in their nest.

199

Chapter 6

Feeding (and Pooing)

In at number five is a majestic big cat – the **African leopard**.

The African leopard is its own enigma. They are shy. Not as shy as the snow leopard, but they are shy. You can follow them – slowly – but you follow them by observing what's happening around you. The alarms of baboons and mourning doves tell you where the leopard is.

There will be certain kopjes, which are rocky outcrops sticking up above the flat billiard table of the savanna, that are favoured by the leopards. Sometimes they're held by lions as their favourite perching spot. The leopards use them as a great place to give birth, where they're somewhat protected. Lions usually give birth under a bush.

How do we film them? Lots of patience. Lots of guides on the savanna all talking to each other over the radio, and an advance party of rangers going off very early in the morning, before sunrise, to the kopjes, favourite trees and den sites, checking those out to find the leopard. Then we can set off with the film crew and equipment. It's all about having a lot of people on deck with local knowledge, so we can cover a lot of ground, because the leopards are so elusive.

Leopards, like most felines, will snooze. Once you've found one, you might have to wait up to eight hours before it does anything. But it's worth the wait. I once filmed them doing this elegant jump from one tree to another. They are so good at climbing they can chase monkeys up trees, force them to make a risky jump from branch to branch and then leap up and catch the monkey as it jumps.

I love leopards. They're so powerful. Like a featherweight fighter – small compared to heavyweight champions like the lions or tigers, but they pack a punch. Their body-to-weight strength-ratio is incredible. They're the gymnasts of the feline world. If you watch them leap down from a height you can see their shock absorbers all bending perfectly, then they're back up again.

Leopards can not only leap upwards to catch prey, they can pounce downwards too. A solitary leopard will lie up in a tree, blending in, and jump down on an impala or a bushbaby, for example. Then once it's caught the animal, it has to carry it back up the tree to eat it. The animal might be twice the bodyweight of the leopard and it has to carry it in its mouth and climb using its claws. Imagine the neck and claw strength it needs to do that! It must take the food up into the tree to keep it away from hyenas and lions (two of the leopard's enemies).

The leopard is probably Africa's most lonely animal. All the other African big cats live in groups, but not the leopard. It wants to keep itself to itself. This does mean it's vulnerable to attack though – which is why it has developed the ability to hide up in trees or in kopjes.

Bagheera in *The Jungle Book* is a black version of the leopard. These tend to be more common among Indian leopards than African.

Facts

Scientific name:	*Panthera pardus*
Mammal family:	cats
Height:	71 cm to shoulder
Length:	160–230 cm
Found in:	woodlands, savannas, forests, mountains, deserts
Eats:	mainly antelopes, gazelles, deer, primates and domestic livestock
Babies:	2–3 per litter
My three words:	elusive, stealthy, elegant

Leopards have big green eyes for hunting at night.

Leopard spots are called rosettes. The pattern gives them good camouflage, but they also have another colour signal. If they've been seen, they lift up their tail and show the little white bit at the end. It's like the white flag of surrender. This means that the baboons – one of their worst enemies – don't need to attack them.

Leopards are solitary, but they communicate with each other using scent, poo and scratch marks on trees. They also call with a cough-like roar, growl when they're angry and purr when they're happy.

Leopards belong to a group of cats called big cats. These are the cats that can truly roar (although the snow leopard is an exception). Other African wild cats such as cheetahs, servals, African wildcats and caracals don't count as big cats.

Predators

Leopards are perfectly adapted to find and catch their food. In fact this is true for all animals – their bodies are designed just right for eating what they need to.

You probably know that animals are known as carnivores if they eat meat, herbivores if they eat plants and omnivores if they eat both animals and plants. Most carnivores are also predators – which means they hunt and kill the animals they eat.

Leopards and all other cats are extremely effective predators. They all have . . .

- **The ability to move very stealthily when they're stalking their prey.**
- **Good eyesight, hearing and sense of touch.**
- **Sharp claws to hold on to prey and for climbing.**
- **Incredible strength and speed so that they can chase and hold their prey.**
- **Excellent camouflage for hiding.**
- **Very strong jaws and sharp teeth for crushing bone.**

Predators come in all shapes and sizes, from small birds, spiders and beetles that eat even smaller animals, to the biggest, strongest, most powerful beasts such as lions, tigers, bears or hyenas.

Predators which are at the top of the food chain and the strongest in their habitat are called apex predators.

In some UK habitats, the otter is the apex predator. They don't have any natural predators of their own. They have an important role in keeping the numbers of other animals and plants in balance.

Weapons

Nile crocodile

Speed ★

15 km/h swimming, 12 km/h running

Size ★★★★

up to 600 kg

Bite strength ★★★★★

4,000–5,000 psi bite force, up to 100 strong, cone-shaped teeth (not used for attack)

Claws ★★

Super-sense ★★★ great hearing

Special weapon   hides underwater

Great white shark

Speed ★★★   56 km/h

Size ★★★★★   1,800 kg

Bite strength ★★★★★   4,000 psi bite force, 300 teeth

Claws   none

Super-senses ★★★★★   great vision, smell

Special weapon   magnetic sense

Psi means pounds per square inch. It is a measure of pressure so is a way to measure how strong a predator's bite is.

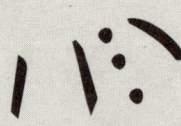

Cheetah

Speed ★★★★ — 105 km/h and very fast acceleration

Size ★★ — up to 64 kg

Bite strength ★★★ — 475 psi bite force

Claws ★★★★ — claws like an athlete's running spikes

Super-sense ★★★★ — excellent vision

Special weapons — big lungs and a very bendy spine to help with sprinting

Polar bear

Speed ★★★ — 40 km/h

Size ★★★★ — up to 800 kg

Bite strength ★★★★ — 1,200 psi bite force

Claws ★★★★ — long, sharp claws for grabbing prey

Super-sense ★★★★ — can smell prey 1 km away

Special weapon — punches holes in the ice to catch seals

Golden eagle

Speed ★★★★★ — 240 km/h dive

Size ★★ — up to 6 kg

Bite strength ★★★ — 700 psi bite force and has a sharp hooked beak

Claws ★★★★★ — 450 psi grip

Super-sense ★★★★★ — exceptional eyesight

Special weapon — kills quickly by puncturing an animal's lungs

Jumping spider

Speed ★ — 3 km/h

Size — around 1 g

Jumping strength ★★★★★ — can jump to 30 times its own height

Claws — (not used for grip)

Super-sense ★★★★★ — superb eyesight

Special weapons — sticky silk, 3D planning skills, venom, trickery

Predators versus scavengers

Being a predator is hard work. Many carnivores make life easier for themselves by eating meat they don't have to hunt, which might be the leftovers from another predator's kill, easy pickings such as the young, sick and weak, or dead meat, called carrion.

Examples of scavengers:

- **Vultures** have bald heads and necks, which stop bits of rotting meat sticking to them.
- **Hyenas** steal meat killed by other animals and feed on any dead carcasses.
- **Crabs** and lobsters feed on dead animal scraps.
- **Crows** often eat roadkill.
- **Bears**, including polar bears, sometimes scavenge leftover human food from bins.
- **Foxes** like leftovers from animal kills and human meals.

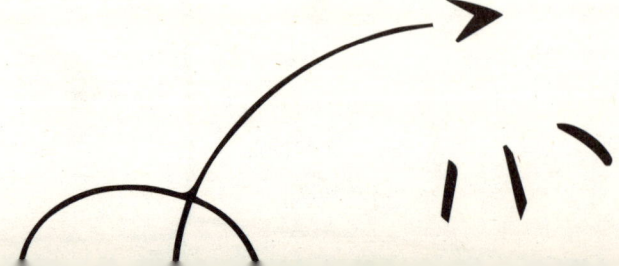

- **White-tailed eagles** scavenge more often than they hunt.
- Loads of invertebrates, including **cockroaches**, **carrion beetles**, **blowflies** and some **ants**, are scavengers, feeding on dead animals or scraps.

Most of these animals hunt as well as scavenge – **hyenas**, for example, can be successful pack hunters.

Hunting the hunters

When filming predators, you have to get into their mindset and think like a predator. After all, you are tracking and following them, and you want to 'catch' them on camera.

So, how do we find them? Sometimes they have collars with tags, but others we find because we get to know their territory and behaviour and know where they are likely to be. It's like humans having their favourite spot on the sofa!

When following predators, you can learn some surprising things. We often go to film polar bears on foot. We carry rifles to defend ourselves but of course we don't want to use them on these precious animals, so we also carry rocks to protect ourselves. Why? The polar bears are scared of rocks. *What?*

If you want to scare a polar bear, all you have to do is clap two rocks together and make that cracking sound. What else sounds like that? Cracking ice, of course. The polar bears have been taught by their mothers that this sound means danger and they have to move away from it, otherwise they might fall through into the water. Swimming uses far more energy than walking, so polar bears would prefer to avoid plunging into the water. But in fact, if they have to swim they are very strong swimmers. They can happily swim for 200 kilometres – which is like swimming the English Channel six times!

So clapping rocks means 'move on, bear'. Ingenious. This is an **aural stimulus**, which means a sound trigger.

If we're in a polar bear's habitat, it's the smell that gives us away. Polar bears hunt by scent and they have an amazing sense of smell. They can sniff a a baby seal 3 metres under the ice, hiding in its den. They can also sense a seal swimming under the ice and they know the seal will have to come up to breathe. So they can wait – they might sit at a breathing hole for hours, just waiting – or they can jump up and down on the ice, trying to break it.

Find out more: Polar bear cubs, page 284

Feeding detective – owl pellets

Examining an owl pellet is a great way of finding out about these hunters. You can put the bits of skeleton that are found in the pellet back together to work out exactly what an owl has had for dinner. It's like a dinosaur dig in miniature – but instead of an enormous creature that died millions of years ago, you're discovering a small animal that was eaten a few days ago.

How to dissect an owl pellet

◎ Get yourself a tray and make sure you're wearing gloves.

◎ Soak the owl pellet in water. When it's soft, very gently (using tweezers or your fingers) pull the bits apart.

Did you know?

Owl pellets are all the bits of an owl's meal that it can't digest – like bones, feathers and fur – coughed up a few hours after it's eaten. You'll find pellets near where the owl sleeps.

- Carefully pick out the bones and lay them out on the tray.

- Look for the main bones – like the skulls, jawbones or perhaps beaks. This will tell you if you've got one animal or more.

- You can identify what the animal is and piece together the whole skeleton using a dissection guide.

Now you're an owl detective! You're finding out all about what food is available for owls in that area. To make your investigation even more special, you can wait until it's dark, and find a good spot to sit and watch the owls as they set off for their night of hunting.

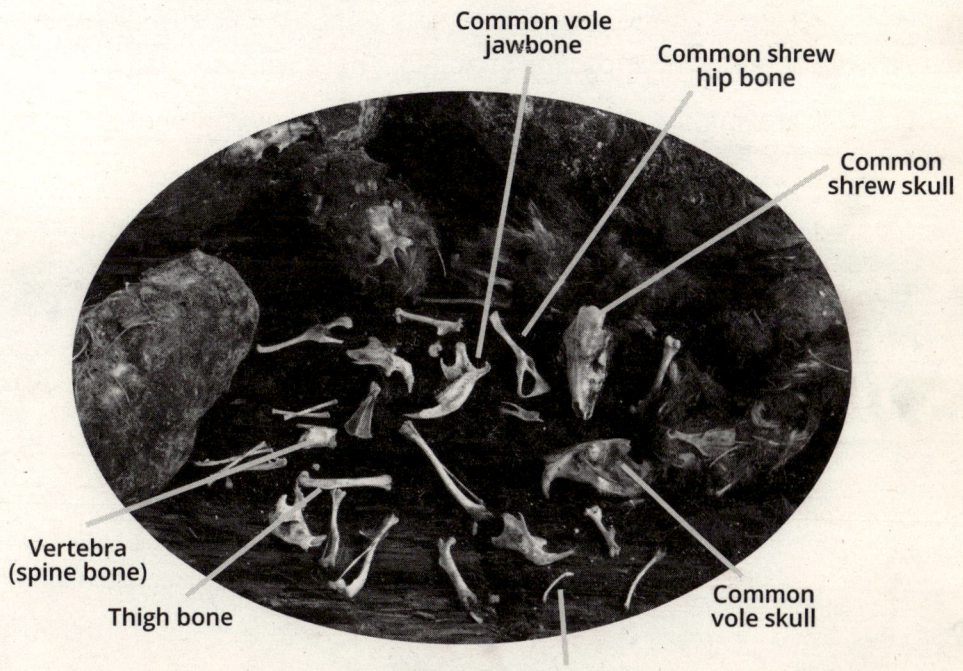

Common vole jawbone

Common shrew hip bone

Common shrew skull

Vertebra (spine bone)

Thigh bone

Common vole skull

Rib bone

The return of the Yellowstone wolves

Yellowstone National Park in the northwest of the USA is the scene of a successful reintroduction project involving a magnificent predator – the grey wolf. It's a brilliant example of how a top predator has a huge impact on everything in its environment.

Grey wolves disappeared from Yellowstone in the 1920s, when humans killed the last pack.

Over time the landscape changed. There were so many elk grazing the shrubs and trees that the environment was becoming less green and was drying out. All the edges of the rivers were worn away. All the vegetation was overgrazed and the birds were gone.

People started to think that there were too many elk and that they needed to be controlled. Maybe by poison, maybe by hunting or trapping? But some biologists and

park rangers, including a man called John Weaver, had the idea of bringing wolves back to Yellowstone.

In 1995, 14 wolves were brought in from Jasper National Park, Alberta, Canada – there were nine males and five females, from a few different packs. They were kept in very large pens for a few days while they got used to their new home, and then they were let out to explore. Each pack established their own territory and soon the animals were moving around their own large areas throughout the year.

Within a year of the wolves' arrival, changes began to happen. The rivers were all straight because the elk and

bison had come right down to the river edges and eaten all the vegetation. Now they could settle back into curving meanders. River meanders provide clear, shallow pools for salmon and areas of riverside plants where insects and birds can feed, as well as slow the river down. The elk and bison hooves had scuffed up lots of silt, so the rivers were sludgy instead of clear, which means it's not so good for the fish.

Immediately, the biologists started seeing fewer elk and bison near the rivers. They had moved up to the higher ground where there are more trees to hide in and it's harder for wolves to catch them. They would only come down to the river once a day to drink, but the rest of the time they'd keep away. So they would not sit there chewing up all the vegetation and churning up the ground too much with their hooves.

This started allowing the vegetation to grow a bit more near the rivers. More grass and shrubs grew. With that, more birds appeared. Some of the birds buried seeds, so trees start to grow. The elk were kept away by the wolves, so the saplings didn't get nibbled.

These changes were all because of one creature: the wolf. The wolf changed the river courses in this corner of North America.

It's a great example of an apex predator controlling everything else in the environment and shows how important it is to have a top predator. We call this a keystone species, because like a keystone in an arch, it holds the whole system together.

Find out more: Keystone species, page 76

Herbivores

Herbivores eat plants. Some eat a wide variety of foods while others are specialists that are very picky about their diet. For example:

Giant Panda

bamboo

Rabbit

grass and other green leaves

Orangutan

fruit, leaves and bark (but also honey and insects)

Locust

acacia tree leaves and buds

Giraffe

Snail

any vegetation

Koala — eucalyptus leaves

Tortoise — mainly leaves

Elephant — grasses, leaves, shrubs, twigs and fruit

Finch — seeds and fruit

Swan — mostly vegetation

Butterfly — nectar

Some adult insects eat nothing at all! They eat when they are larvae, and as adults they focus all their energy on finding a mate and laying eggs.

Guess the feeding style from the skull

Here are some skulls I've collected and cleaned. Looking at an animal's skull tells you a lot about what and how they eat. The answers are at the bottom of the next page.

1 A slender, pointed beak for probing and picking its food, which consists of spiders, insects, earthworms, seeds, grain and fruit. This one still has its tongue. It was killed by a barn owl.

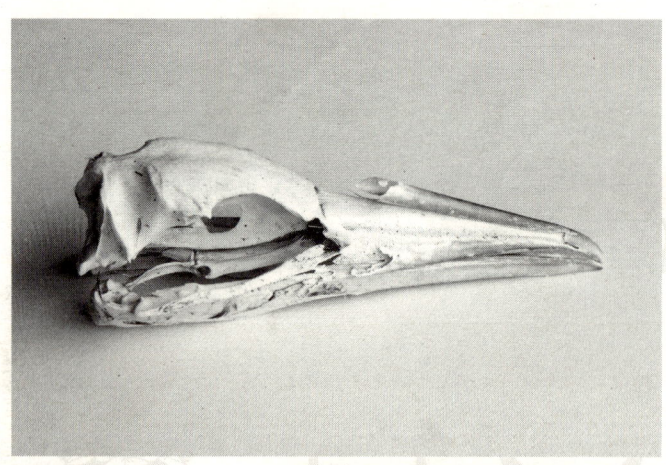

2 A very long, fine beak. If you touch it you will feel it's actually bendy at the tip. The bird can move the tip of its bill around in the mud or underwater to feel for insects and worms. The tip is very sensitive to touch. Its bendy nature means that the bird can open and close the end of its bill under the mud to grab its prey. This skill has a special name: **rhynchokinesis**.

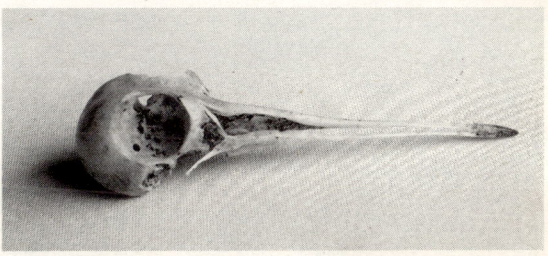

3 Sharp incisors for biting off pieces of meat, jagged back teeth for holding slippery fish and strong lower jaws for crushing food. Notice that its nostrils are high up on its head so it can breathe while swimming along.

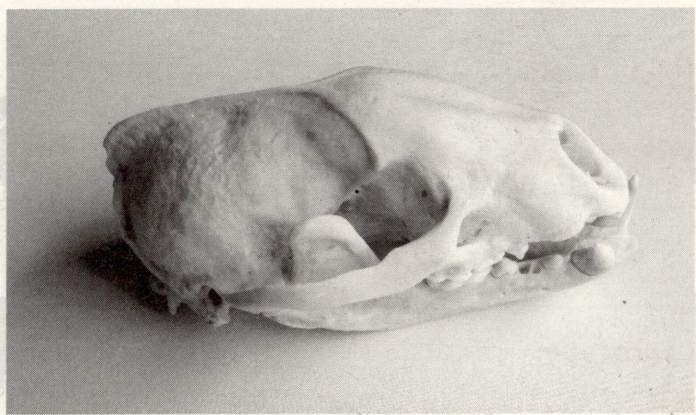

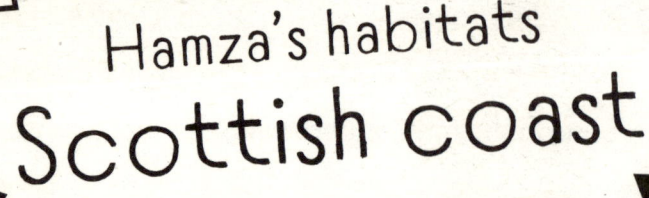

Hamza's habitats
Scottish coast

Habitat:	Scottish coastal habitat
Where in the world:	around the coasts of Scotland's mainland and islands – more than 18,000 km of coastline in total
Landscape:	coastal – very varied

One of my favourite habitats is the coastal habitat right outside my house. It's one of the best things ever. It's a place where two worlds collide. Land animals mix with sea animals and then you get animals that live in between the land and the sea, such as seals. Everywhere you go is full of life and there is enough food to support a huge variety of animals.

I can see seals from the front of my house and a golden eagle from the back, believe it or not. On a typical day I can see harbour seals, otters, red deer, white-tailed eagles, a golden eagle, curlews, oystercatchers, mink, dolphins, minke whales, six or seven types of geese, shelducks,

stoats, weasels, corncrakes, orcas, wigeon, teal, red-breasted merganser, and sheep grazing the seaweed. And that's without even examining the sand and rock of the beach and all the small animals that feed there.

Scotland's shores are especially rich for wildlife because of their position in the world, where the warm Gulf Stream currents and cold polar currents mix, and because there is such a long coastline and large expanses of sea. There are many different habitats within the coastal habitat. There are . . .

- Sandy beaches and sand dunes, home to seabirds, waders, otters, seals, dolphins, porpoises, crabs and mussels and other rock pool creatures.
- Salt marshes, where thousands of waders, ducks, geese and swans feed, home to rare flowers and butterflies.
- Salt lagoons, where the saltiness of the water is in between seawater and freshwater, an unusual habitat home to rare plants and a feeding ground for ducks, geese, swans and waders.

◎ Rugged clifftops, home to colonies of nesting seabirds.

◎ Machair, a unique coastal grassland habitat found only on Scotland's Western Isles and the western coasts of Ireland, incredibly rich in wild flowers and home to waders, corncrakes and rare bees.

These habitats are very precious. If there's a disaster on land or at sea – such as an oil leak, run-off from a chemical spillage or sewage overflow – it's the coast that gets hit first.

Nature jobs

If you want to work directly with animals, getting close to them and looking after them, a zoo or wildlife park is one place to do it. There are lots of different jobs in a zoo. Here are a few:

I'm a zookeeper – I take care of the animals every day. I get pretty close to the animals, but there's a lot of clearing up poo!

I work in conservation, rearing young animals to be strong and fit and hopefully to one day be released into the wild.

I am a zoologist. I study the animals that live in the zoo – I can make much more detailed observations than I'd be able to do in the wild.

I'm a vet – I look after the animals if they are sick, as well as giving them their injections and other treatments, just like a doctor. I need to know about a wide range of animals, from lions and tigers to snakes, fish and small mammals.

I'm the education curator, helping the public to understand the animals and the work of the zoo. Most days I meet a school group.

Animals in zoos are ambassadors for their species. They're educational and they're a way for people to see animals and understand more about them. Many zoos allow young people to be a zookeeper for a day.

Decomposers

Herbivores eat plants. Carnivores eat herbivores. And some carnivores eat other carnivores. Nutrients and energy flow from food to feeder. But it doesn't end there. The nutrients and energy get recycled and doing this is the job of the scavengers, detritovores and decomposers.

Scavengers and detritivores eat dead things and decomposers break them down so they can be recycled and provide goodness for more plants and animals. Fungi are some of the most important decomposers, but lots of bacteria can do this, too.

Some eat dead animals and bits of animals, while others

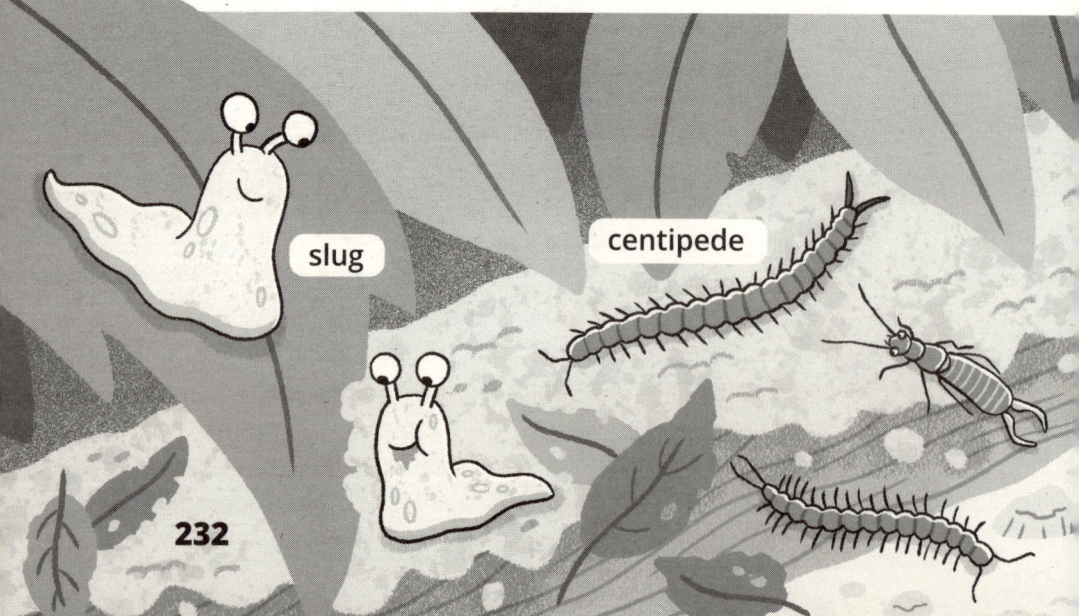

slug

centipede

eat dead and decaying plants, such as fallen leaves and wood. There will be beetles and beetle grubs inside the log, eating away at the wood.

Be sure to put back any logs or stones that you look under – the animals need the damp, dark conditions.

Another important group of recycling heroes are animals that eat poo! Dung beetles are superstars at this. The adults eat dung and lay their eggs in it, and the eggs hatch into grubs that feed on it, too. Several flies also lay their eggs on the dung for their grubs to eat. Without these decomposers, tonnes of animal poo would never disappear – and the world would be covered in poo!

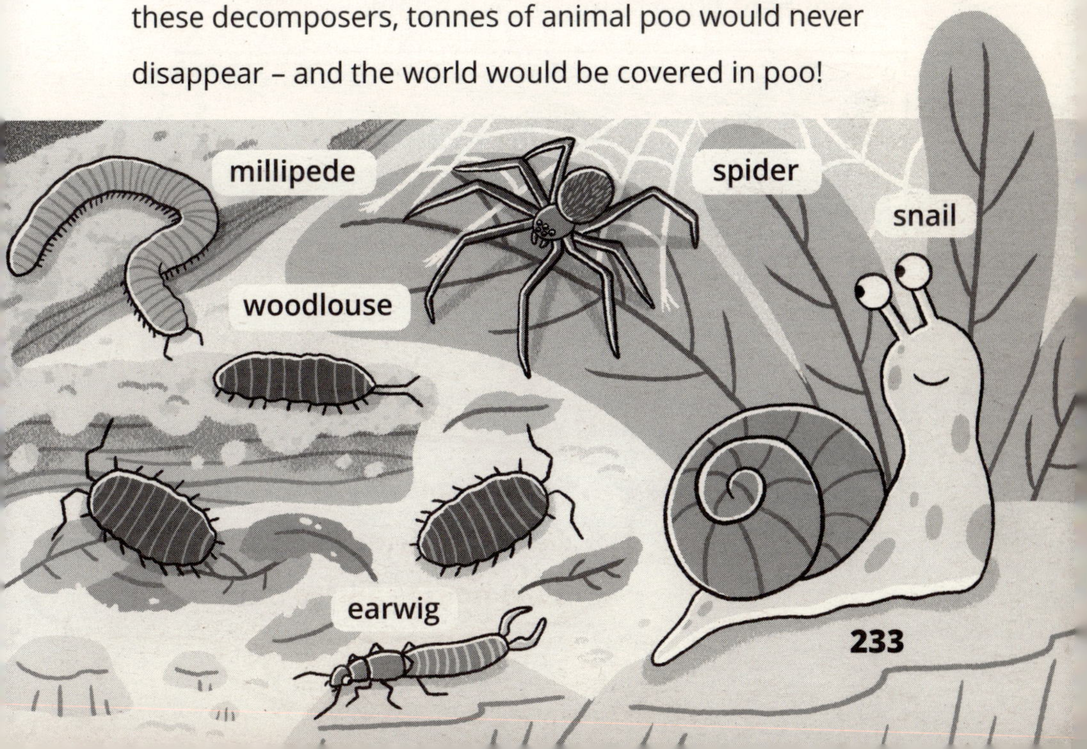

millipede

spider

snail

woodlouse

earwig

Poo!

Poo may be gross but it's fascinating and important. It's the major way that animals get rid of undigested waste from the food they eat, but it does lots of other jobs, too. It can be . . .

- ◎ **A nursery** – dung beetles and dung flies lay eggs in poo.
- ◎ **A building material** – many birds use dung as one of the materials for building and lining their nest. Oriental skylarks use whole balls of elephant dung to make their home!

Even to this day, most human hunter-gatherer societies use animal dung to build houses and to burn as fuel.

◎ **Fertilizer** – poo makes an important fertilizer for adding nutrients and roughage to the soil so it's full of goodness for plants to grow.

◎ **A communication tool** – poo is full of visual and smelly messages for other animals of the same or different species.

◎ **A seed-spreader** – hundreds of different flowers rely on animals to spread their seeds by eating them and pooing them out.

◎ **A cooler** – some types of stork poo on their own legs to help them cool down!

You, too, can pick up some of the messages in animal poo. The simplest one says which animal has been there and left that poo.

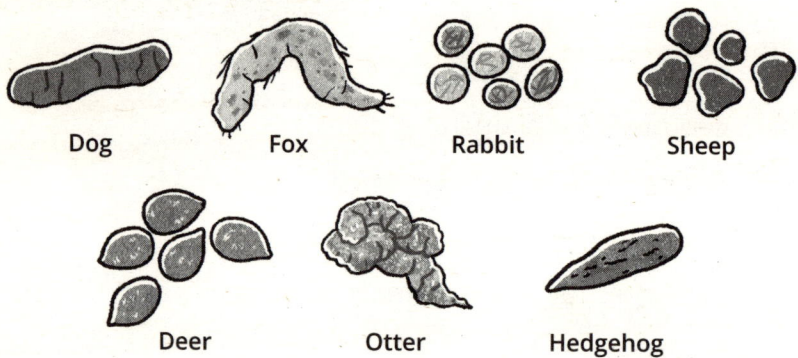

| Dog | Fox | Rabbit | Sheep |

| Deer | Otter | Hedgehog |

Remember, never touch animal poo – but you can poke it with a stick to have a better look at what's in it.

We train dogs and cats to poo in certain places. Remarkably, I managed to train a pine marten to poo in a box, like a cat in a litter tray!

I have pine martens living in my attic. They were pooing on my attic hatch. I would open the hatch and the poo would fall on me. Pine martens like to poo on the same spot. They build up something like a poo volcano! That

marks their territory and says 'This is my home – keep out.' The size of the volcano and the fact that there is fresh poo on top tells other pine martens how long it has been in that spot.

I can't have them pooing on my hatch because I need to get in and out. I want to keep them in my attic, though, because I want to film them. So what I do is get the poo and put it in a box with newspaper in the bottom. Every two days I go and collect the layer of paper with the poo and throw it out. Basically, it's a litter tray! I trained pine martens to use a fish box as a litter tray. If you take the poo away, they want to make sure that spot still has their scent, so they'll keeping pooing there rather than all over the corners of the attic.

In the poo you can see what they're eating – if they're eating berries, birds or other animals.

10 Animal Poo Facts

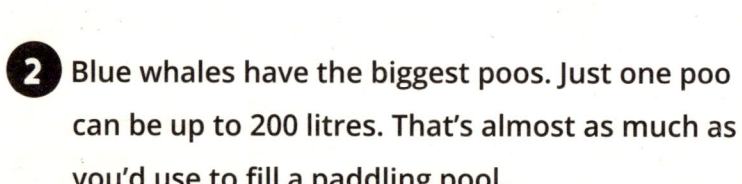

1 Sloths only poo once a week. They climb all the way down from their tree and poo at the base of it, before climbing back up again. It's the only time they really move anywhere.

2 Blue whales have the biggest poos. Just one poo can be up to 200 litres. That's almost as much as you'd use to fill a paddling pool.

3 Rabbits eat their poo. In fact, many mammal herbivores do the same.

4 Elephant and hippo poo can be made into paper.

5 Some caterpillars shoot poo far away from their bodies so that they don't attract predators.

6 Parrotfish poo turns into white sand. Whole beaches are made up of it.

7 Bat poo (or guano) has been used to make fireworks and gunpowder.

8 Turkey vultures' poo contains good bacteria that kills off the bad bacteria found in their food (dead meat). So they clean their feet by stepping in their own poo.

9 Emperor penguin poo-stains on the ice of Antarctica help scientists find where the penguins live and breed.

10 A wombat's poos are cube-shaped! Lots of animals send messages with their poo, but only the wombat does it with a tower of little cubes.

Hamza's Nature Heroes
Steve Irwin

1962–2006

When I first saw Steve Irwin on TV, wrestling a crocodile, I couldn't believe it. At home in Sudan, everyone knows that you never go after a crocodile. There are loads of myths about how dangerous they are and you always know that they might be lurking in the Nile or other rivers.

So imagine my thoughts when I turned on the TV and saw this dude wrestling a croc. I just thought, 'One of these days a crocodile is going to eat you!' But the excitement and danger was irresistible.

Steve Irwin was an Australian conservationist, TV presenter and educator. His *Crocodile Hunter* series saw him get dangerously close to fierce and deadly animals, such as venomous snakes and spiders, sharks, Komodo dragons, hippos and lots of alligators and crocodiles. He wanted to stop crocodile hunting in Australia and helped move crocs to sanctuaries where they would be safe.

240

In 2006, Steve had a terrible accident. He was filming near Australia's Great Barrier Reef when he swam near a huge bull stingray. The animal attacked him and punctured his heart. Tragically, he died.

His passion for wildlife, especially endangered wildlife, was communicated to people everywhere and his family have continued his conservation work.

Of course, I understand the thrill of getting close to dangerous animals in the wild. I've experienced it myself when filming polar bears, big cats and even birds of prey. It's always important to remember you're in the presence of something dangerous, and to respect the wildlife. Hopefully some of that excitement comes across on-screen and you can feel it too when you watch my films.

Chapter 7
Communication

We're getting towards the top of my top ten now, and coming in at number four we have another big cat: the **snow leopard**. These secretive cats are known as the 'ghosts'. They live in the Himalayas and they're incredibly difficult to find.

These 'ghosts' hunt ibex. The ibex are phenomenal because they can walk on practically vertical cliffs. They have to be amazingly agile because of how incredible a hunter the snow leopard is.

A snow leopard can sit in plain sight and it'll be camouflaged. There could be one sitting right in front of you and you wouldn't see it until it moved. Their speciality is running down the mountain to hunt ibex. They use gravity to do this – and this is where their tail comes in handy. They have a super-long thick tail compared with their body, and they use it like an acrobat on a tightrope uses a pole – for balance.

Ibex are difficult to catch, so snow leopards also go for the next easiest prey – domestic sheep, cattle and yaks. They're much easier to bring down.

Snow leopards are big, fluffy animals – and that's because of the terrain, to keep them warm. Their paws are slightly bigger, relative to their size, compared with other cats. This is to help with grip but also means the paws act like a snowshoe, providing warmth and spreading the cat's weight to stop it sinking into the snow.

Out of all the cats in the world, apart from the clouded leopard, the snow leopard is one of the toughest to find and film. The terrain is so harsh – you need to be altitude trained or go through acclimatization to get there. They're basically the top predators on top of the world, not that far from Mount Everest. Not many people see them, let alone film them.

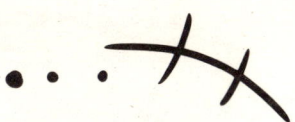

The best way for a filmmaker to track down a snow leopard in the white expanse of the mountains is to use a camera trap (a camera with a motion sensor that is triggered when an animal walks past it). But where do you put it? The clue is in the wee . . . You need to find their spraint sites, which are scrapes in the ground. This is how they communicate. They'll paw and scrape the ground, then lift their tail and spray wee – telling everyone this is their territory.

Mark Smith was the guy who spent three months trying to film them for the BBC's *Planet Earth* series – even through Christmas. He managed to get two minutes of an outline of a snow leopard prowling across a mountain ridge, super-far away. That's three months' work. That's how much of a ghost they are.

I love them. Hopefully one day someone will send me out there to film them.

> **Unlike other big cats, snow leopards can't roar. Instead they make a piercing yowl.**

Facts

Scientific name:	*Panthera uncia*
Mammal family:	cats
Height:	60 cm to shoulder
Length:	60–150 cm, plus tail 80–100 cm
Found in:	steep, rocky, snowy mountains
Eats:	ibex, sheep, goats, yak, marmots, hares
Babies:	usually 2–3 per litter
My three words:	ghost, fluffy, elusive

Another name for the snow leopard is ounce.

Snow leopards can jump up to 9 metres from rock to rock.

Snow leopards' tails are very soft and fluffy, so they're also good wrapped around the cat's body for extra warmth.

In the cat family, the snow leopard has the largest tail compared to its body size.

tiger   lion   leopard

panther   snow leopard   lynx

Communication

Animals may not talk using words, like we do, but they have a wealth of precise and effective ways to communicate with each other. They use . . .

Visual communication

– colours, patterns, body language (including dancing!), sometimes facial expressions

Sound – a huge variety of calls, as well as using other objects, for example hitting a tree or drumming the ground

Smell – using poo, wee, sweat or other chemicals to make different smelly messages

Taste – communicating warnings such as 'You don't want to eat me!'

Touch – communicating through cuddles and grooming or through pushing and shoving during a confrontation

Find out more: Dancing, page 296; calls, page 258; birdsong, page 262; smelly messages, page 280

251

Camouflage

You probably know that camouflage means using colours and patterns to blend into the surroundings. In a way it's the opposite of communication – using visual information to be noticed as little as possible instead of to draw attention to yourself.

Lots of animals use camouflage to hide in plain sight. A toad hiding on a leafy woodland floor, a tawny owl blending into the bark of an old tree and a grasshopper perching on a blade of green grass are all using colours and patterns – and the act of staying still – to avoid being seen by predators or prey.

A tiger seems brightly coloured to us, with its handsome black-and-orange stripy coat. But most of their prey can

only see in shades of grey, so the tiger blends in against the long grass of the jungle. It's similar with zebras. A single zebra standing still may blend in with long grass. In a whole herd of zebra, the black-and-white stripes confuse predators. A lion can see a stripy jumble of zebras but it's hard for it to pick out the shape of one individual zebra, so it's hard to know where to strike.

Lots of animals have spots for camouflage. Think of the leopard and cheetah – or, closer to home, the starling, fallow deer and song thrush.

Some animals change their colour at different times of year so they stay camouflaged whatever the surroundings. In the Arctic, Arctic foxes, snowy owls, mountain hares, weasels, lemmings and

ptarmigan (a kind of grouse) all grow snow-white fur or feathers in winter and become brownish in summer.

Many animals make sure their eggs or babies are camouflaged. Red deer have plain reddish-brown coats, but their calves (babies) have spotty coats that help them blend into long, flowery grass. The calves don't give off a smell and know straight away to keep still and quiet so they're not giving out any signals to predators.

Birds that lay their eggs on the ground, such as lapwings, skylarks, plovers and Sandwich terns, lay speckled eggs that blend in perfectly with pebbles, grass or straw.

Colour

Animals can use colour to stand out as well as to hide away.

Bright colours and patterns are a way for animals to recognize each other. A butterfly's bright wings or a bird's colourful feathers are like a name badge. They also communicate information about how strong and fit they are and whether they will make a good mate.

Some animals put on impressive visual displays to show they are ready to mate. A male ruff (a kind of wading bird) grows an elaborate, colourful ruff of feathers around its neck during the breeding season.

Mandrills (a type of large monkey found in Africa) use bright colours to show dominance. Big males intimidate each other

I'm blue in the face.

by showing off their bright red-and-blue faces. They can impress other mandrills enough to get them to back off from a fight.

Chameleons use colour to communicate their mood. They have the amazing ability to change colour – not to blend in with their surroundings, as many people think, but to react to cold or warmth and also to signal that they are calm, anxious or angry and ready to fight.

White-tailed deer lift their tail and flash the white patch underneath as an alarm signal to warn other deer that danger is nearby and it's time to start running.

Birds' courtship colours and displays, pages 292–4

Sound

Animal calls – called vocalizations – communicate a range of pieces of information.

I'll tell you where the food is . . .

I'm scared!

Where's my baby?

Let's play!

Sound, amazing examples

Animals do some extraordinary things with sound – using their voices and other body parts.

◎ **Male gorillas** make an 'oo-oo-ooo' call with their mouths at the same time as beating on their chest. They cup their hands under their pectoral muscles and use their chest like a soundbox. The bigger the male, the louder and lower the sound will be – so he's showing he's boss and impressing the females.

◎ **Pistol shrimps** use sound as a deadly weapon. They use their strong claw to produce an explosive blast of bubbles which creates a shockwave so loud that it stuns the shrimp's prey.

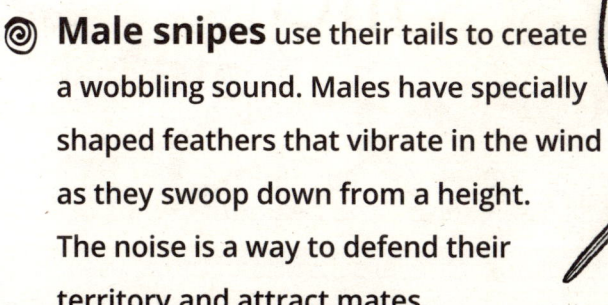

◎ **Male snipes** use their tails to create a wobbling sound. Males have specially shaped feathers that vibrate in the wind as they swoop down from a height. The noise is a way to defend their territory and attract mates.

◎ **Male lions** roar so loudly that the noise shakes the ground. The sound is deep and low-pitched and can be heard for miles. Hearing a lion's roar is an awe-inspiring experience: you feel it vibrating through your body and are in no doubt about its power.

◎ Some kinds of **snakes** in hot countries make a rasping sound by rubbing their scales against each other. This is a better option than hissing, which means opening the mouth and losing precious moisture.

◎ Some kinds of **bumblebee** make a special type of buzz that vibrates at exactly the right frequency to make certain flowers release their pollen. Ordinary buzzing won't do the trick.

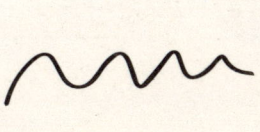

Birdsong

One of the most interesting – and beautiful – kinds of animal call is birdsong. One large group of birds, the songbirds or passerines (which I call tweety birds), are the experts at this musical skill.

In the northern hemisphere, it's mostly male songbirds that sing. They sing to defend their territory, show off how strong and fit they are, and attract a mate.

Zoologists are finding out more and more about female birds' songs around the world. In Australia, the female fairy wren sings to compete with other females and attract males. More than 60 per cent of the world's female tweety birds sing.

Learning to recognize birdsong makes your birding even more enjoyable. Sometimes, it's the call that allows you to identify a bird, especially if you only get a quick glimpse. There are lots of apps and websites you can use to learn different birds' songs. My favourite is Merlin.

Ten songs

Here are some UK birdsongs to learn and enjoy. Have a listen online.

- **Blackbird** – melodic orchestra
- **Robin** – clear and beautiful
- **Tawny owl** – 'twit twoo', sung by two different birds
- **Song thrush** – repeats itself three or four times then changes the song
- **Woodpigeon** – 'coo cooo coo'
- **Wren** – a trill at the end
- **Chiffchaff** – 'chiff chaff'
- **Nightingale** – melodic techno
- **Yellowhammer** – 'a little bit of bread no cheese'
- **Skylark** – melodic, endless singing from the heavens

Robins are very territorial birds. Females and males both sing to defend their territory all year round.

Male and female swallows sing to each other, and sing different songs.

Find out more: Dawn chorus, page 340

Body language

We use body language all the time – communicating using facial expressions and the way we move our body. For example, being 'wide-eyed' means excited and alert, 'squaring up' means making yourself big and threatening and getting close to someone as if you're about to fight. Someone who's slouching back in their chair and looking at the floor is probably uninterested or uncomfortable with the conversation and someone who's fidgeting a lot might be anxious. Smiling is one of the most understandable examples of body language, but even a smile can mean different things.

Body language is our animal language! Other animals use it, too – although animals' gestures might not always mean the same as ours. In many (though not all) human societies it seems friendly and honest to look someone straight in the eye, but to dogs – and gorillas – this is aggressive.

m m . :

You probably know a lot of dog body language. Think of a dog with hunched shoulders, head down and tail between its legs – it's probably frightened. If it's wagging its tail and panting, it's happy and excited. If it's baring its teeth and has its tail sticking straight up, it might be angry.

If bears don't want to fight, they walk away, sit down, yawn and generally act like they are ignoring the other animal. If they want to show dominance and frighten the other animal, they stand up on their back legs, call out, slap the ground and even charge.

- Apes and monkeys use a lot of facial communications.
- Prey animals such as zebra and deer prick up their ears when they are frightened.
- Rabbits drum their feet on the ground when they are alarmed.
- Many animals use body language to signal to each other that they are ready to mate or to impress a mate.

Animals talking

Who's a pretty boy?

Animals may not talk using words, like we do . . . or do they? A few animals have learned human language.

Parrots can learn to copy human words and phrases. We don't think they fully understand their meaning but they do understand some of the information that goes with them. They can learn that people say a particular phrase in a particular situation, such as 'Nice to see you' when someone walks into the room. They can learn words for colours, shapes and food, and understand combinations of those words to get the meaning.

Chimpanzees can learn human sign language and have been able to have conversations with people they've

got to know. In the 1970s, keeper Kat Beech taught sign language to a chimp called Washoe. Kat was away for a while because she had a baby who sadly died. When she eventually came back, Washoe seemed annoyed with her, so Kat signed 'My baby died'. The chimp made the sign for 'cry' and gently touched Kat's cheek, showing she understood.

I have had the amazing experience of talking with a chimp. At the Welsh Mountain Zoo there was a chimp called Tuppence, who had learned sign language. One day she pointed to my fizzy drink and to the gap in the enclosure fence, signalling that she wanted me to pour out some drink for her to have. Then she kept doing a gesture a bit like rubbing her hands. I asked the keeper what it meant and they said it means 'Me and you are friends.' Tuppence was asking if she could be my friend! Then she signed to say she wanted to play the tickling game Round and Round the Garden. Phenomenal! I felt so lucky and privileged.

Warnings!

Animals need to be able to communicate a warning to others if danger is near or to ward off a predator that's getting too close.

Danger! Danger!

Beavers slap their tails on the water as a sign of danger.

Many butterflies, such as the peacock, flash big eye-spots if a predator comes near. The predator is fooled into thinking these are the eyes of a bigger animal like an owl or a snake.

Some moths reveal a flash of bright colour to startle and confuse a predator. Yellow underwing moths are generally brown but flash the bright yellow underside of their wings when threatened.

Some octopuses flash bright blue to warn predators not to risk being on the receiving end of their venomous bite.

Don't messssss with me!

Cobras display their hood as a warning. They don't want to use up resources and energy attacking something with their venomous bite – they would rather just scare the other animal away.

Many animals use colour as a warning that they are poisonous, taste bad or have a nasty sting. Black and white, bright orange, yellow and red are common warning colours.

Mimicry

Lots of animals use **mimicry** – or copying – to send messages. Many butterflies, bugs, frogs and snakes that are neither poisonous nor venomous mimic the warning colours of those that are, to trick predators into steering clear.

Milk snakes – black and red bands, yellow stripes – mimic **coral snakes**.

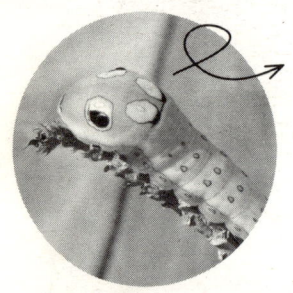

The **spicebush swallowtail butterfly caterpillar** – green with yellow eyespots – mimics the **smooth green snake**.

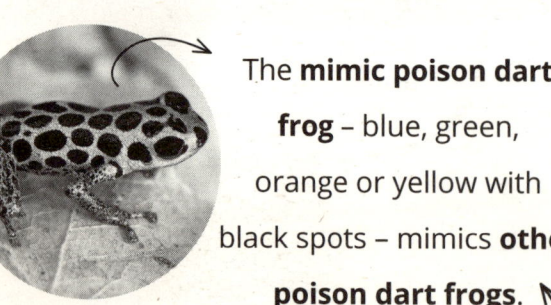

The **mimic poison dart frog** – blue, green, orange or yellow with black spots – mimics **other poison dart frogs**.

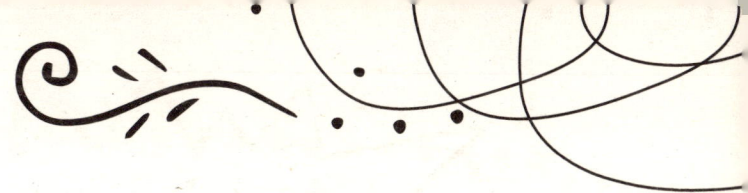

Many animals use mimicry for camouflage to go a step further than blending into the background, by disguising themselves as something in particular. The buff-tip moth looks exactly like a broken-off piece of birch twig. The Vietnamese mossy frog looks just like a clump of green moss. The leafy seadragon looks like floating seaweed.

Lots of birds are sound mimics. They copy the calls of other birds, other animals and even human noises such as phones, gates and alarms.

Remember which is a coral snake and milk snake with the rhyme 'Red on yellow, kill a fellow. Red on black, friendly Jack.'

10 Stripy Insects

These insects communicate using colour, saying 'Go away', 'Don't eat me', 'Look out, I'm poisonous' and 'Careful, I sting'. Some of them are poisonous or do sting, but others are harmless mimics using warning colours.

Honeybee
Black and gold

White-tailed bumblebee
Black, gold and white

Wasp
Black and yellow

Rainbow leaf beetle
Metallic green, blue, gold and red

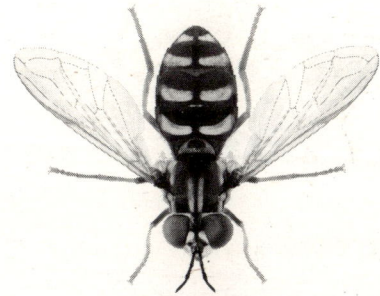

Hoverfly
Black and gold

Rosemary beetle
Metallic green and purple

Wasp beetle
Black and yellow

Ten-lined June beetle
Black and white

Colorado potato beetle
Black and gold

Glorious jewel scarab
Black and green

Nature jobs

There are lots of things you can do if you want to use your love of nature in creative jobs, like me.

I'm a zoologist, wildlife cameraman and wildlife presenter. I film wildlife and make programmes that tell people all about it.

I am a wildlife film producer. I think up ideas of what I want to make a film about, organize shoots and work with the camera people.

I'm a film editor. It's my job to cut, shape and polish hours of footage into spectacular programmes of the right length that people will love.

I'm a script writer. I write the script that the wildlife narrator will read. Sometimes a famous actor reads it but sometimes it's a wildlife expert.

I'm an artist. I paint wildlife and wildlife scenes.

I'm an author. I write books about nature for children.

I'm a teacher and I love teaching my students about nature and helping them to love it, too.

Hamza's Nature Heroes
Dian Fossey

1932–1985

Great apes are some of my absolute favourite animals, so of course the zoologists who led the way in studying them are my heroes. For gorillas, the leading zoologist was Dian Fossey.

Dian Fossey went to eastern Africa in the 1960s and worked for Louis Leakey, a famous anthropologist (someone who studies people and cultures, often in the distant past). Leakey was studying fossils of the earliest humans. He thought that it was important to make detailed observations of the lives of the great apes – our closest relatives – as a way to understand more about human evolution.

Leakey thought that women would make better observers than men. They would be more patient, and pay more attention to the females, babies and relationships instead of focusing on males and dominance. Dian was one of the women he hired to study primates (the animal group that apes, monkeys and lemurs belong to).

Dian set up a research centre in Rwanda's Virunga region, the jungle home of the mountain gorilla. She lived there for many years, patiently observing the gorillas and finding out about their habits, communication and relationships.

In the 1970s one of her favourite gorillas, called Digit, was killed by poachers, or illegal hunters, and this spurred Dian on to tell the world about poaching and the need to protect gorillas. Tragically, her passionate support for the gorillas ended up costing her her life. In 1985, her body was found near her campsite. It is thought that she had been killed by poachers.

Even though her work style was controversial to some people, the discoveries she made were invaluable. My friend Vianet Djenguet is one of the people building on her work – he's just made a fantastic film called *Silverback*.

Find out more: Jane Goodall, page 318; gorillas, page 364

277

Courtship signals

Animals' behaviour leading up to mating is called courtship. They signal to impress potential partners. They do this using colour, calls, body language, smell or taste – and usually a combination of these.

Visual

Glow-worms and fireflies (actually types of beetle) signal that they're ready to mate by putting on a light display. They make a glow-in-the-dark chemical in their bodies and switch on the glow to help potential mates find them. With glow-worms it is only the female that glows, whereas with fireflies both sexes light up, with hundreds of them flashing in answer to one another.

The peacock shows off with his magnificent tail.

Calls

Male mice signal they are looking for a mate by making an extra-high-pitched whistling song.

Female tigers roar and call out to tell males they are ready to mate.

Body language

Male black widow spiders do a very energetic bottom-shaking dance when approaching the female, so she won't mistake them for prey.

Smell or taste

Male adders follow scent trails left by females.

Some animals go to great lengths, beyond simple signals. The male Japanese puffer fish creates a beautiful sand sculpture to impress females. It is a perfect circle with ridges and valleys and it takes him seven days to build. If it is a success, the female will lay her eggs in the centre.

Find out more: bird courtship dances, page 296

Smelly messages

Many animals are far more sensitive to smell than we are, and use smell to send all sorts of messages. They do this in their wee, poo and in other chemicals they give off, sometimes in oil from special glands.

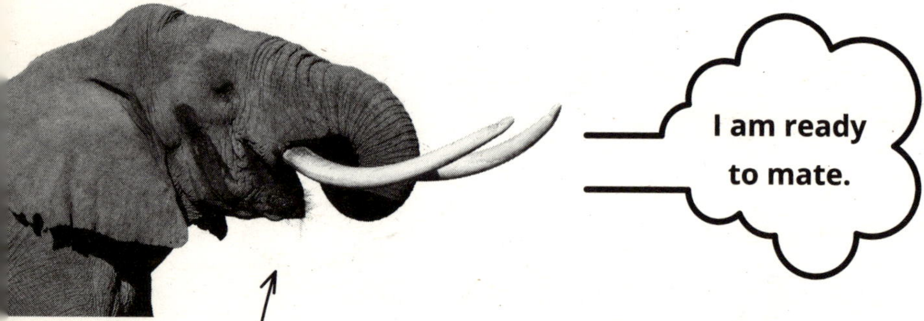

I am ready to mate.

Bull elephants in musth (ready to mate) secrete a substance from glands on the side of their head. It's one of the smelliest chemicals there is.

Some male moths have such sensitive antennae that they can smell females' scent from 3 kilometres away.

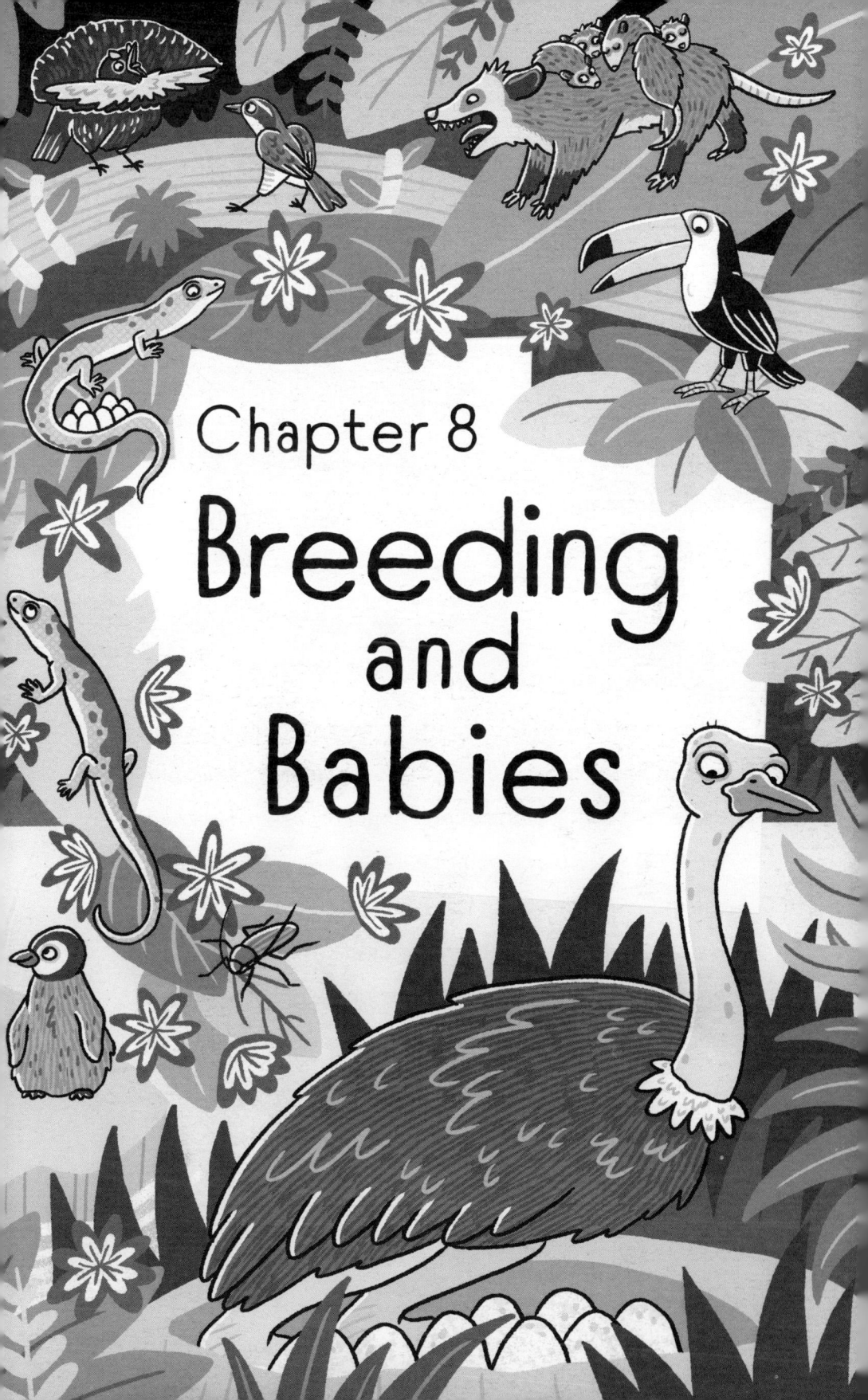

Chapter 8

Breeding and Babies

Now, let's get back to my top ten countdown. Coming in at number three is the **polar bear**. It's the best bear!

I've been lucky enough to film polar bears in Churchill in Canada. One time we were filming a mum and cubs. I'm busy with the drone, watching some mini icebergs floating along in one direction. Then our guide pipes up with, 'Guys, polar bear!' OK, so I was watching the mum and cubs, but the guide points out that there is an iceberg going in the other direction from the rest, and this is also a polar bear! The whole of his body is underwater, except for his head. He is about 20 metres away – a young juvenile. They are the trickiest ones to deal with, the teenagers. He starts coming straight for us. So we have to stand up in a line and say, 'Right, we're

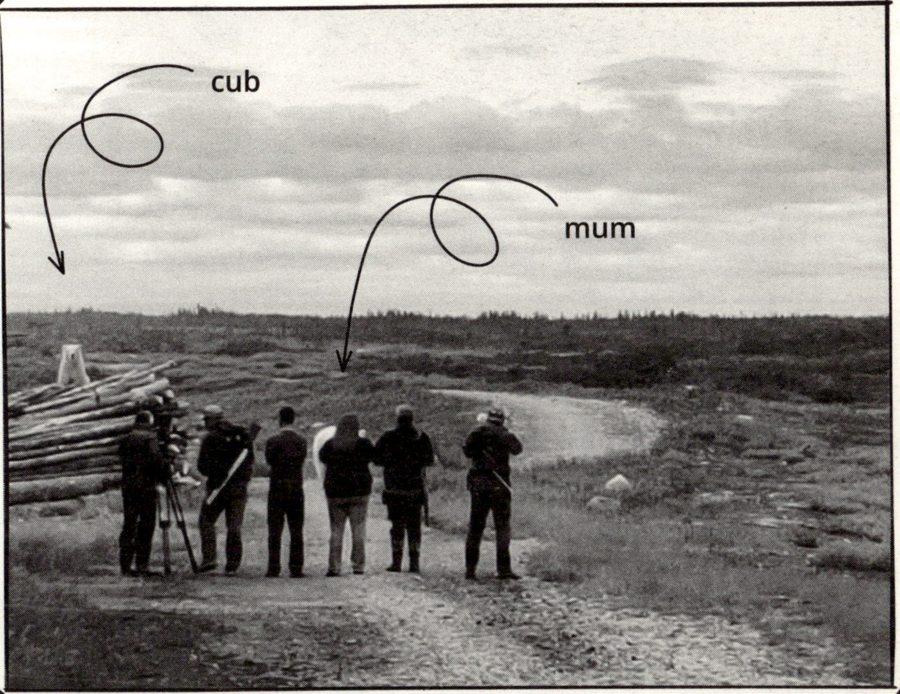

cub

mum

bigger than you.' One guy gets on top of a rock to make
himself even bigger.

I'm thinking, my drone is far away, I need to get it back.
But we couldn't figure out what the bear was interested
in us for. What we didn't realize was that the wind was
blowing out to sea away from us – but our quad bike,
which contained all our food and so on, was behind us. So
our scent and the food scent was wafting straight to him.
He's thinking, 'That's an easy meal!'

So I bring the drone back towards us, and as I do so he looks up at it. The guide says, 'Hamza, arc around us and then send it as far away as possible.' So I do, and the bear follows the drone and we move quickly, get to safety, and finally return the drone.

me lying in a polar bear day bed

I got to film a really playful cub who was completely cool with us and wanted to play on these wood piles. He's just nursed, so he's got a foamy milk moustache round his mouth and he's full of beans. The mum is not so happy about this. Mum knows to leave humans alone, but she'll also protect her cub. So he wants to play and he wants to come close to us, but we have to be really careful not to get too close and anger the mum. Like any mother bear, she is at her most dangerous if she thinks her cubs are threatened. For example, if he's playing and he falls or gets his foot jammed or something, and he squeaks, we're in trouble.

Facts

Scientific name:	*Ursus maritimus*
Mammal family:	bears
Height:	up to 1.6 m tall to shoulder
Length:	up to 2.5 m
Found in:	Arctic landscapes
Eats:	ringed seals, harp seals, hooded seals, whales, walruses, seabirds and eggs, small mammals, fish and carrion (dead meat)
Babies:	1–4 (usually 2) per litter
My three words:	fierce, swimmer, giant

Polar bears especially like blubber, a thick layer of fat that whales and seals have, because it is so high in energy.

Young polar bears in their first year are called coys (Cubs Of this Year) or yearlings.

Polar bear feet are brilliant for swimming. They are massive and webbed like paddles. They're good on land, too – their paws spread out like snowshoes to spread their weight so they can walk on thin ice without breaking it, and tiny bumps for grip and long claws both stop them sliding on the ice.

The polar bear is the largest land carnivore on Earth.

Baby size

Giant panda · Cub

adult 1.5 to 1.8 m
baby about the size of a
domestic cat's kittens

Orca · Calf

adult up to 2.4 m
baby 2.6 m

Leopard · Cub

adult up to 230 cm
baby up to 15 cm

Gorilla · Infant

adult up to 185 cm
baby 40 cm

Red kangaroo   Joey

adult up to 1.6 m tall, with tail 1.1 m
baby 2.5 cm

African elephant   Calf

adult up to 4 m tall
baby up to 1 m tall

Polar bear   Cub

adult up to 250 cm long, 160 cm tall
baby 30 cm long

Green Turtle   Hatchling

adult up to 120 cm long
baby 5 cm long

Getting together

For animals, the most important thing they can do in life is to breed, passing their genes on to the next generation. To give their young the best start in life animals need to be fit and strong to breed, choose a good partner of the opposite sex, and come together with that partner to mate and breed.

The behaviour that leads up to mating, as we talked about in the communication chapter, is called courtship. Different animals do this in different ways. Many animals call, sing, dance or display themselves. Some fight, to show off how strong they are. Others build nests, to show they will be a good parent.

Some give presents:

- Male penguins give pebbles to their partner as a romantic gift that also helps her with nest-building.

- A male South American spider called **Paratrechalea ornata** gives the female a present of a prey insect wrapped up in silk. The better the prey, the more impressed she will be.

- Toucans pass a gift of food back and forth to each other until the female accepts one once and for all.

A bird called the grey shrike catches prey and stores it impaled on thorns. To impress a female, the male displays his thorny 'larder' and the female will choose a male with the most impressive food store.

The male bowerbird builds a phenomenal construction called a bower and decorates it with bright, colourful items such as rocks, shells and berries, or even coins and small bits of coloured plastic. This is a stage for his courtship dances to impress the female. She then makes her own simple cup-shaped nest to lay her eggs in.

Mating itself happens in all sorts of weird and wonderful ways. Some animals, for example albatrosses and beavers, mate with just one partner and generally stay with them for their whole life. In others, one male mates with many females (red deer, gorilla) or a female mates with many males (phalaropes, bees), while still others mate with many different partners every year. Some fish pair up, but some don't even meet their mates – the female releases eggs into the water and a male releases a cloud of sperm to fertilize the eggs.

Find out more: bird courtship dances, page 296; mammal mums, page 320; bird parents, page 324

Courtship dances

Great crested grebe

1. Shake your head and show off your crest.
2. Bend your neck backwards like a curtsy.
3. Dive under the water and leap back up again while your partner ruffles their wings.
4. Dive for some weed and swish it back and forth, standing tall on the water.

9

Superb bird of paradise

1. Start with a bow.
2. Flash your eyes from blue to yellow.
3. Step from side to side, whirr your wings, shake your head feathers.
4. For a final flourish, flash your iridescent throat patch.

10

Mute swan

1. Face your partner, lift your wings and bow gracefully.
2. Make a loveheart shape with your necks.
3. Bow, move away and make the loveheart shape again.
4. Keep that beautiful wing position.

Greater flamingo

1. Stand tall and shake your head like a flag.
2. March around with your beak held high.
3. Flick a wing out to the side and flash those bright feathers.
4. Mix up your wing moves to stand out in this group dance.

Sharp-tailed grouse

1. Bow your head and stick up your tail.
2. Hold out your wings.
3. Stamp your feet – tappity tappity tap – and spin in a circle.
4. Beatbox while you dance and push the other dancers away.

Laysan albatross

1. Bow to your partner and bob up and down.
2. Shake your heads, clatter your bill and kiss your partner.
3. Rise up as tall as you can, bill to the sky.
4. Stretch your bill under your wing.

Jackson's widowbird

1. Jump.

2. Jump.

3. Jump.

4. Jump – as high as you can!

Red-capped manakin

1. Fly up from your perch.

2. Swoop in and do a loop-the-loop in the air.

3. Land, lift and quiver your tail.

4. Hold on to your sides.

5. Step back and slide – it's a moonwalk.

It's a ten from me!

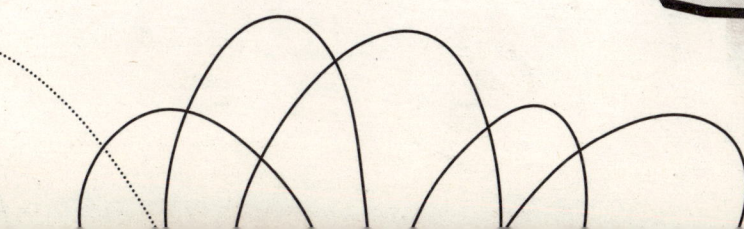

Nests

Many animals build nests to lay their eggs or have their babies in and as a safe place to look after them when they're very young. There are some great nest-builders . . .

1. Weaverbird – father builds a hanging basket nest out of grass.

2. Harvest mouse – mother weaves a perfectly round, tennis ball-sized nest from grass.

3. Rhinoceros hornbill – both parent birds work together to peck out a nest hole in an old tree trunk, before the mother hops inside to lay her eggs and the father plasters the entrance up until it's just a tiny hole he can use to feed her. She's safe but trapped until she pecks her way out when the chicks are ready to fly away.

300

4. Betta fish – father makes a bubble nest using saliva (spit), as a safe place for his fertilized eggs, then stands guard over them.

5. White stork – makes one of the messiest bird nests, a huge pile of sticks full of droppings, food remains such as bits of bugs and even puddles of water.

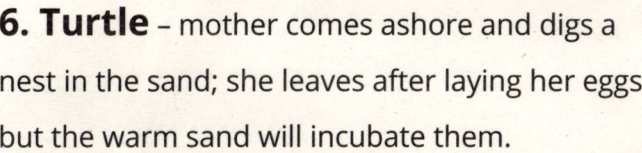

6. Turtle – mother comes ashore and digs a nest in the sand; she leaves after laying her eggs but the warm sand will incubate them.

7. Polar bear – mother digs an underground den that is up to 25 degrees warmer than the temperature outside.

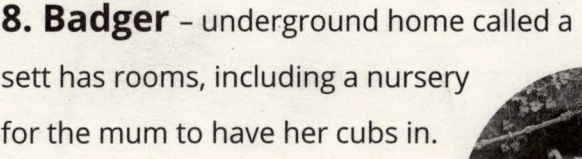

8. Badger – underground home called a sett has rooms, including a nursery for the mum to have her cubs in.

Nests in detail

Osprey

Where: in a tall treetop or on a cliff or high platform

Made of: sticks

Lined with: bark, grass, algae

Built by: both parents, with the male fetching most of the material and the female doing most of the building

Size: up to 180 cm across and 390 cm deep

Amazing fact: ospreys often nest within about a mile of each other, because if they can see a successful nest, they know it's a good place

Ringed plover

Where: bare ground on pebbly or sandy beaches

Made of: nothing – just a shallow dent in the ground

Lined with: nothing

Built by: the male

Size: about 10 cm diameter

Amazing fact: if a predator approaches, a parent bird will run away and flap around pretending to be injured so the predator will follow it instead of taking a chick

Kingfisher

Where: not a nest but a tunnel inside a river bank

Made of: hollowed out of the mud

Lined with: regurgitated fish bones and scales

Built by: both parents

Size: up to 140 cm long and just 5 cm wide

Amazing fact: the egg chamber in the tunnel is angled slightly so eggs can't fall out. As the chicks grow, the nest fills with stinky rotting leftover fish bones – yuk!

Long-tailed tit

Where: often in a thorny bush or shrub
Made of: moss and lichens woven together with spiders' webs
Lined with: up to a thousand soft feathers
Built by: both parents
Size: about 15 cm deep and 12 cm wide
Amazing fact: as the chicks grow, the nest grows! The springy nature of the moss and the stretchy spiders' webs allow it to expand

Swallow

Where: often in a building such as a barn
Made of: mud mixed with grass
Lined with: grass and feathers
Built by: both parents
Size: about 7.5 cm across and 5 cm deep
Amazing fact: it takes a pair of swallows about 1,200 trips to gather the little pellets of mud they use to build the nest

You shouldn't get close to a bird's nest because you don't want to frighten the parent birds or draw attention to the nest and help predators find it. But some naturalists set up nest cams and you can enjoy watching with these. Or sometimes you might find an old nest in the winter after the bird has finished with it, and then it's fine to have a good look.

305

Nest boxes

One way I can get some fantastic film footage of bird behaviour is to set up a nest cam inside a nest box. I can see and film in detail how the parents take care of their eggs and chicks, and I can watch the chicks grow. Nest boxes are a great way to give birds a helping hand, too.

You can buy or make nest boxes for lots of types of birds, or make them from recycled materials.

Classic – for blue tits, coal tits, great tits, house sparrows

Open – for robins, wrens

Owl box

I make my own barn owl nest boxes from old barrels and hang them from the roofs of buildings so that pine martens can't reach them. I line them with crushed-up owl pellets. This is what the owls themselves use – they are full of fur and feathers and make lovely soft lining material.

You can also get nest boxes for bats, red squirrels and dormice.

In the summer I provide water to make mud puddles for the swallows to get their nesting material.

Swifts nest in the eaves of houses and other buildings. In the last 25 years, more than half of our swifts have disappeared and it's partly because new buildings don't have any holes for them. But builders can use special swift bricks or house owners can put up swift boxes.

Some tips for your nest box . . .

◎ Find a good site – about 2.5 to 5 metres off the ground, angled forward slightly so rain won't get in, somewhere between north and east and not in direct sunlight.

◎ Make sure it is well away from cats and foxes – it should not be somewhere they can easily climb up to.

- Make sure the bird has a clear flight path in and out of the nest, without obstructions in the way.

- Don't put it too close to a bird table or bird bath as all that activity nearby may put breeding birds off.

- Ideally, put the box in position in autumn – this means the birds can get used to it and make their mind up about it over winter, and also gives them an opportunity to roost there when it's cold.

- Each autumn, ask an adult to help you brush the inside of the box clean and wash it with boiling water so that you don't get germs building up.

Eggs

In nearly all animals, the female lays eggs. Eggs protect the developing young creature inside and are full of goodness that gives it a start in life. Frogs and fish lay thousands of eggs at a time and sea urchins and ants lay millions, whereas some penguins and albatrosses lay just one egg each year.

Some animals don't lay eggs. Their eggs develop inside the mother's body and she gives birth to live young. This is the case for nearly all mammals. Some snakes and sharks have live babies and there is a toad, the Suriname toad, that lets her babies develop under a patch of skin on her body before they burst out of a hole in her back!

There are two types of mammals that DO lay eggs. They come from the most ancient surviving group of mammals and are found only in Australia and New Guinea. They are the platypus and the echidnas.

Egg champions

The biggest egg belongs to the ostrich. It's 15 centimetres long and weighs 1.3 kilograms (about as heavy as 24 chicken eggs). But the ostrich is also the biggest bird. The kiwi is only about the size of a chicken, but its egg is still 12 centimetres long.

The tiny bee hummingbird lays the smallest bird's egg – it's only about 12 millimetres long. But many fish eggs, frogs' and toads' eggs, insect eggs, spider eggs and other invertebrate eggs are far smaller than this.

Egg shapes	Pointy	Round	Oval	Long
Birds	guillemot, razorbill	ostrich, owl	thrush, blackbird	
Reptiles		turtle	crocodile	lizard, snake

Find out more: platypus and echidna, page 322

You can see huge numbers of razorbills at their breeding sites on the UK's coasts.

Razorbills . . .

- Nest in groups of between 2,000 and over 20,000 birds. The nest site is a **noisy, smelly** place with all those birds croaking and growling and pooing! There are often thousands of puffins, guillemots and kittiwakes as neighbours, too.

- Lay their eggs high on **cliffs**. They don't make a nest but choose a spot on a ledge (but not too close to the edge!) or in a crack in the rock.

- Sometimes get **aggressive** towards other razorbills as they all jostle for space. They shout at each other or clash bills.

- Lay one egg each year. The egg is cream-coloured with black spots and squiggles. Each one is different, which helps the parents to recognize their own egg. It is also pointed, which means it will just roll in a circle if knocked, rather than rolling away.

◎ Incubate the egg for 35 days, with both parents taking turns to sit on the egg.

◎ Spend 18 days feeding their chick. Both parents work hard to catch enough sand eels, cod and herring. While one parent is finding food, the other stays with the chick to keep it warm.

When the chicks leave the nest, they jump off their ledge and plop down to the sea. The parents stay close to the jumplings and feed them for a few more weeks before they can find food for themselves.

Hamza's habitats
Arctic and Antarctic

Habitat: polar regions

Where in the world: the Arctic and Antarctica

Landscape: The Arctic is frozen treeless land surrounding the partly frozen Arctic Ocean; Antarctica is an enormous continent of ice surrounded by the near-freezing Southern Ocean

Another of my favourite habitats is at the Arctic and Antarctic – well, it's really two habitats, at the ends of the Earth.

I've filmed in the Canadian Arctic – home of the mighty polar bear. There's something primeval about being next to the world's largest land predator and knowing that you are on their menu – although hopefully not right then and there!

The oceans in the polar regions have some of the richest marine life of any waters in the world. Cold water has more oxygen than warm water, and also allows nutrients from the depths (some of this comes from whale poo) to rise up to where plankton can use them. This kicks off a food chain that in the Arctic includes krill, cod, herring, jellyfish, bowhead whales, beluga whales, narwhals, walruses, Arctic terns, seals, orcas and polar bears, among others.

In the Antarctic I've filmed in the South Sandwich Islands. This wild, uninhabited place is home to king penguins, wandering albatross, terns and many other birds.

It's an incredible landscape. Your mind doesn't let you understand how big a glacier really is until you see it. You can get 11-kilometre-long glaciers that you would think were solid land because you've been sailing alongside them for a full day. It wraps your brain around how big things are, and also the terrifying speed at which we are changing the landscape of these ice worlds. Seeing it makes you care more than ever.

Did you know?

◎ The Arctic tern migrates from the Arctic Circle to Antarctica – and back – every year, experiencing summer at both poles. In its lifetime, one Arctic tern flies a distance that is the equivalent of flying to the Moon and back.

◎ Polar bears never meet penguins – polar bears live up north in the Arctic whereas penguins live down south in the Antarctic and some other places in the southern hemisphere.

Rockhopper penguins and macaroni penguins sometimes breed with each other. The offspring are called rockeronis or macahoppers!

317

Hamza's Nature Heroes
Jane Goodall
Born 1934

When I was lucky enough to study chimpanzee behaviour at the Welsh Mountain Zoo, I was building on the work of chimpanzee expert Jane Goodall. As a young woman, Jane had the chance to travel to east Africa and work with Louis Leakey, the anthropologist who also worked with gorilla expert Dian Fossey.

She set up camp in Tanzania, where she spent months studying chimpanzees in their natural habitat. She observed their family structure and social life as well as all their eating and sleeping habits.

She was the first scientist to realize that they eat both meat and plant food, and the first to realize they use tools. For example, Jane observed that they eat leaves, then dip them in water. They don't want to drink water with muck in, they're using the leaves like a sponge. They chew them up then soak them in water and squeeze them out in their mouths.

Usually scientists keep their distance from animals so that they are not influencing their behaviour in any way. But Jane did the opposite – she befriended them and they allowed her to become a low-ranking member of the troop. She gave all the chimps names, got to know them as individuals with their own personalities and learned a lot about their family and community dynamics. She has been studying them ever since – for 60 years.

In the 1970s she set up the Jane Goodall Institute to support research and conservation of chimpanzees and their habitats. Like all the wild great apes, chimpanzees are declining because of habitat loss and hunting. The institute helps local people live and work in harmony with the forests and nature. It also runs a sanctuary to care for chimps rescued from illegal hunting.

Find out more: Dian Fossey, page 276; chimpanzees, page 176

Mammal mums

In nearly all mammals, mums give birth to live young and feed them on milk made by the mum's body. Milk is full of the nutrients and energy that babies need. Different animals have different amounts of fat, sugar and protein in their milk. Seals' milk is the richest and fattiest – more like cream!

 Giraffe mums don't sleep at all for several weeks while their calves are young.

 Mother elephants produce around 11 litres of milk a day – that's over 40 mugfuls.

 Bat mums roost together in big nursery colonies, separate from the males. They leave their newborn pups together in clusters called crèches while they fly off catching food.

- Bear mums will fiercely protect their cubs from any danger or threat and will even chase away big, strong male bears that get too close.
- Dolphin mums sing lullabies to their calves.
- Koala mums chew their own poo, regurgitate it and feed it to their joeys. This is because the joey's stomach is not developed enough to digest eucalyptus leaves (koalas' only food) and they need to have it pre-digested by the mum.
- Koalas, along with kangaroos, wombats, possums and wallabies, belong to a group of mammals called marsupials. They give birth to young that are so tiny and underdeveloped that they can't survive outside the mother's body. So the mother keeps them in a pouch where they do some of their growing.
- A rabbit mum might have 360 babies in her lifetime.

Rule breakers

Echidnas and the duck-billed platypuses are rule-breaking mammals because they lay eggs. They are the only members of a group of mammals called monotremes. The platypus is found only in Australia, while the four species of echidna live in both Australia and New Guinea.

They both . . .

- Lay eggs, which hatch into babies called puggles.
- Feed their young with mother's milk which the puggle suckles from special hairs, not teats like other mammals.
- Have a beak and no teeth.

Echidnas . . .

- Are small spiny animals that look like hedgehogs.

- Lay a single egg and then keep the egg in a pouch until it hatches and for a few weeks afterwards.

- Can curl up into a ball to escape danger.

The platypus . . .

- Has waterproof fur and is a great swimmer.

- Uses electrosense along with touch to find its way around underwater and to find prey.

- Lays two eggs and cares for them in a burrow.

- Is one of very few venomous mammals – the male uses venom to compete with other males.

- Gives off a bioluminescent green-blue glow under ultraviolet light.

Bird parents

Birds take care of their eggs and young. They incubate their eggs, which means sitting on them to keep them warm, until they hatch. Then, in more than 90 per cent of bird species, both parents are the carers.

Emperor penguins have a tough parenting job. The dad keeps the single egg warm by holding it on his feet and keeping it warm tucked under the feathers of a special flap of skin called a brood pouch. Meanwhile the penguin mum sets off in search of food. She returns two months later, when the newly hatched chick is ready for its meal.

Some birds sing to their eggs before they hatch. The birds learn to recognize the mum's voice, and they sometimes whistle back.

Laysan albatrosses are devoted parents. For the first few weeks, they take it in turns, one staying with the chick and the other foraging for food. When the chick is bigger and needs more to eat, they fly off for up to 4,800 kilometres and seventeen days at a time to find food.

The **cuckoo** is famous for not taking care of its own young. Instead she lays an egg in another bird's nest, even matching the colouring of that bird's eggs. The host bird will look after the egg, which hatches before its own chicks do. The chick will even push the other chicks out of the nest – it is much bigger than them so this is not difficult. This strategy allows the cuckoo to lay up to twenty-five eggs in different nests and have another bird do all the work of raising her chicks.

> **Wisdom, a Laysan albatross who nests in Hawaii, is 73 years old and is still able to lay eggs.**

Unusual parents

We expect mammals and birds to care for their young. But these animals do, too.

1. **Earwig** – mother tends to her eggs and grooms them to remove any mould.

2. **Water bug** – father carries masses of fertilized eggs on his back until they hatch.

3. **Wolf spider** – mother spins a silk sac for her eggs and carries them on her back until they hatch.

4. **Giant Pacific octopus** – mother lays hundreds of thousands of eggs, and watches

them, fanning them with water to keep them clean and oxygenated, until they hatch. This takes six months or more – she doesn't eat during this time and she dies when the eggs hatch.

5. **Strawberry poison dart frog** – father wees on the fertilized eggs to keep them moist, and mother carries them one by one on her back all the way from the forest floor to a miniature rainwater pond in the leaves of a plant called a bromeliad. Each egg needs its own miniature pond otherwise the babies will eat each other.

6. **Cichlid fish** – some mother cichlids keep their eggs and larvae safe in their mouths.

7. **Clownfish** – like Nemo's dad, clownfish fathers care for their eggs by guarding them, fanning them and eating any that get damaged or mouldy.

Some parents eat their young! **Blennies** (a type of fish) usually look after their young, but if the number of eggs is not big enough, they can get bored and eat them instead.

The return of the white stork

The white stork is a magnificent bird that has lived alongside humans for centuries. There are many old legends throughout Europe about how storks bring people babies. In some Muslim countries, people revere storks because their migration seems to guide pilgrims to Mecca. In ancient Egypt the stork symbolized the soul.

White storks make messy nests on tall treetops and often on rooftops. But until recently none had nested in Britain since 1416! They would still visit the UK as a detour on their migration, so surely they could be reintroduced?

The chosen location for a project to bring the white stork back to the UK was Knepp in Sussex, site of a phenomenal rewilding project run by Isabella Tree and Charlie Burrell.

From 2016 onwards, white storks were brought to Knepp and kept in a very large, fox-proof, mink-proof pen. Some

Storks were the first birds that made Europeans realize that birds migrate.

were injured birds from Poland (hurt by flying into power lines, for example) that had been rescued and looked after at Warsaw Zoo. They couldn't fly, which meant they would stay put. They were joined by young birds reared in a zoo in the Cotswolds. The hope was that they would be like a magnet for any migrating white storks that happened to be passing, that they'd attract these migrants to stay awhile and even breed. By 2020 some of these birds had started to nest – and the first white storks born in Britain for 600 years appeared. The white stork was back!

Several of the Knepp-born storks have flown away as truly wild birds and have been recorded in southern Spain and Morocco, while others have stayed and thrive in the rich wetland and scrubland habitat that has been created at Knepp.

In 2023 there were 11 nests at Knepp and 26 chicks fledged.

Chapter 9
Day by Day

Nearly at the top of my favourites list, in at number two, is my top bird. It might not be one you'd expect – it's the **starling**, or European starling to be precise. It's an underrated bird, but it's great.

The most phenomenal thing the starling does is the murmuration. This is the word for a flock of starlings, but it's more than that – it's a beautiful display of about 40,000 to 50,000 birds (sometimes as many as a million!) all in the air at the same time, making beautiful swirling shapes in the evening sky.

Scientists have figured out that the murmuration behaviour is to tell everybody the best shared information and experience about food spots. It's also a case of safety in numbers and it's also a hierarchy of thousands of birds, with every bird knowing their place.

A huge flock will attract predators. The more birds there are, the more eyes are going to be watching them – the

eyes of eagles and sparrowhawks and peregrines. All of these are thinking, why hunt one starling when I can hunt 40,000 at once? But for the individual starling it's much less of a risk than facing one of these predators on its own. When the peregrine comes to dive at them it will shoot straight through because it's confused and doesn't know which one to aim for.

The starlings all kind of talk to each other in the flock, but they also precisely monitor seven other starlings around them. None will crash into another. They weave and swirl and change direction, but they all know what those seven are doing so they all twist and turn together elegantly.

The way they do this is they instantly calculate what the majority are doing. So of those seven neighbouring birds, if two go one way and five go the other way, they start moving towards the five, and then the last two will join.

It's a beautiful display to watch. You can sit in among it and feel the power of it. (I mean, 40,000 of them pooing on you is a feeling . . . !)

The starling is a common and familiar bird. Everyone knows it and it's underrated because there are so many of them. It's actually a really beautiful bird, with speckled and shimmering plumage. If you see them around people – for example, at a service station or in a town square or park – you'll see how brave they are, and boisterous, too.

Some starlings live in the eaves of my house. They're very clever. Early in the year they let house sparrows bring in most of the nesting material, start making nests . . . but when things are looking just right for the starlings, before the sparrows can lay their eggs, the starlings kick the sparrows out. So there's a squabble for about two days before the sparrows go and start another nest elsewhere and the starlings go into the eaves and build their nest. Every year this pair do the same – they raise about three or four chicks and then they leave.

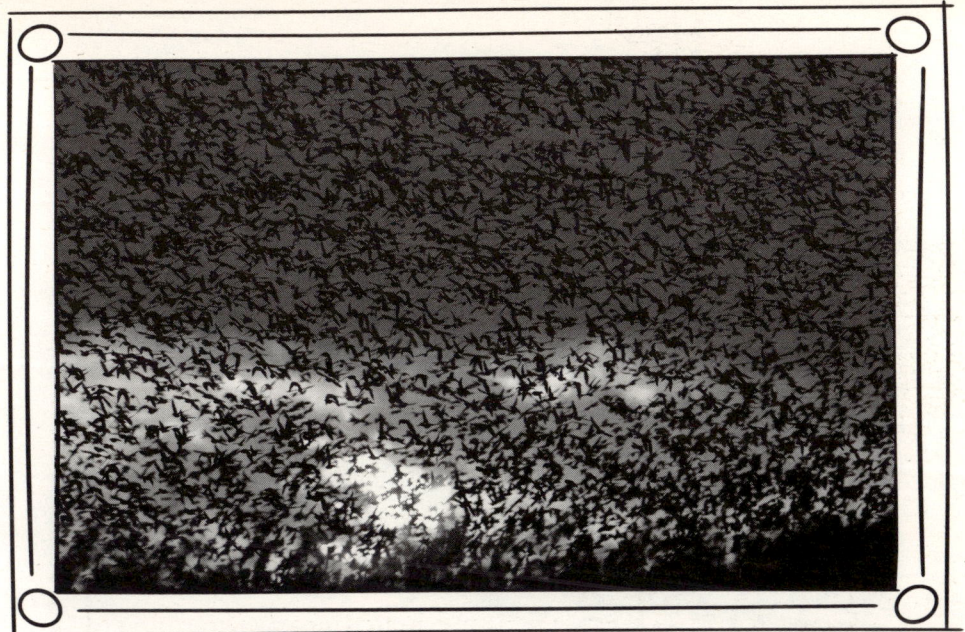

I need to get my fascia boards replaced at the house but the first thing I'll do when the new ones go up is to put a hole in them to say that starlings are welcome. They were here before me; I'm not kicking them out. I'll make sure the work gets done in the autumn, after the breeding season is over.

Another great thing about starlings is that they are mimics. They copy other noises – not just other birds' calls but all sorts of noises, such as dripping water and human noises like those made by phones, alarms and machines. The more they mimic, the more attractive they are to the opposite sex. I have this one starling who likes to perch on my roof and mimic my car alarm. That's pretty cool.

In Rome, Italy, a super-murmuration of around 10 million starlings has been seen swirling above the city in recent winters.

Facts

Scientific name:	*Sternus vulgaris*
Bird family:	starlings and mynas
Length:	21.5 cm
Wingspan:	37–42 cm
Found in:	grassland, parks, gardens, fields, towns
Eggs:	4–5 per clutch
Three words to describe:	cheeky, clever, mimic

Starlings live in the UK all year round, but in winter thousands more of them come here from mainland Europe, so winter murmurations can be vast.

Get a close-up view of a starling and you can see its beautiful plumage. In winter they are black-brown with silvery white speckles and black bills. In summer their bills turn yellow and their feathers become glossy, with an iridescent sheen of blue-green or purple. This is another way of showing off that they are fit and strong and will make a good mate.

Many species of starling have even brighter iridescent colours than our European starling – bright blue, yellow, turquoise and orange.

European starlings were introduced to New York's Central Park in the 1890s, by people who wanted America to have birds mentioned in Shakespeare's works. They spread over the whole of North America and are now a pest there.

Body clocks

Every day we follow a simple pattern. We wake up in the morning, we carry out lots of activities such as eating, playing, going to school or work and many other tasks throughout the day, and we go to sleep at night.

Our bodies know when it is daytime and when it is night-time. We have an internal clock called a body clock, found in a part of the brain called the pituitary gland, which keeps all the cycles in our body moving at the right pace. Our body clocks respond to daylight, which 'resets' the clock each day, meaning we tend to feel awake when it is light and sleepy when it is dark. People who work at night sometimes feel sleepy doing their jobs and find it hard to sleep in the day. When we go on a long flight across several time zones, we feel sleepy or wide awake at odd times until our body clock catches up.

All animals have a body clock and follow daily patterns of activity. Plants do, too – you can see flowers opening and closing, or stems moving throughout the day.

Except in tropical areas, hours of sunlight in a day vary throughout the year. Animals' body clocks are sensitive to these changing light levels.

- Sheep become ready to mate when days get shorter in autumn. This means the lambs will be born in spring, when the weather is warmer and the grass is lush.
- Many birds prepare to migrate when day-length shortens and autumn comes.
- Songbirds sing more often when there is more daylight, in spring.

Find out more: Diurnal and nocturnal animals, page 342

Dawn chorus

Early each morning, just before the sun comes up, tweety birds burst into song. From their perches high in the trees they can see the light peeping above the horizon. The early-morning light triggers them to start singing before the rest of the day's noise and activity really gets underway. Some sing all day, but it is at dawn that they give it their all and really put on a musical spectacular. This early-morning music is called the dawn chorus.

Each day, the earliest singers are blackbirds, robins and song thrushes, who begin when there is hardly any light in the sky. Soon other birds join in. At this time of day,

nocturnal predators are tired and daytime predators are not yet up. It's still too dark to look for food.

The dawn chorus is loudest and most energetic in spring, when songbirds are defending their territories and showing off to attract mates. In the UK, it's most impressive from mid-March to early May.

The dawn chorus is changing. In some places it is not as loud or varied as it once was. Loss of habitat means there are fewer different species of birds singing. On the other hand, in some cities it is starting earlier and some birds are even singing louder, fighting to be heard above the noise pollution.

The dawn chorus is not the only period of birdsong in a day. Many birds have a burst of song or calls in the evening, and a few, such as the nightingale, are famous for singing at night. This is so that they don't have to compete with the other birds' songs. Nightingales collect up to 1,000 musical tunes and phrases from their migration.

Find out more: Birdsong, page 262

Sleep

When an animal is asleep, it is vulnerable to predators. But it must be important for animals to sleep, because nearly all vertebrates, as well as some insects and even worms, do it. We think that sleep allows the body to rest and the brain to sort information.

Animals that are awake in the day and sleep at night are called **diurnal**. Animals that are awake at night and sleep during the day are called **nocturnal**. Animals that are awake mainly at dawn and dusk are called **crepuscular**. This is a good time of day to be awake because animals miss the hottest and coldest times of day, the dim light makes it easier to hide and predators are tired from a night of hunting or only just waking up.

Animals have lots of tricks to make themselves less vulnerable when sleeping. Some hide away in a burrow, others sleep in groups, and some manage with just short power naps.

Dolphins have an amazing sleeping skill. They sleep with just half their brain at a time, with the other half staying awake and alert. If they went into a deep, complete sleep, like we do, they would stop breathing and drown.

Hibernation is a special kind of long sleep that some animals use as a way of coping with winter. The animal makes a cosy warm burrow or den where it hides, safe from the cold and predators. It usually eats as much as possible and gets fat before hibernating. This gives it energy reserves to survive on for many months during the winter. When hibernating, an animal hardly breathes. Its body temperature drops and its heart rate slows right down. Hedgehogs, ground squirrels, chipmunks, bats and dormice all hibernate.

Zzzzzzzzzzz

Did you know?

The koala sleeps for between 20 and 22 hours every day!

Roosts

Roost is the word for bird or bat sleep. It also means a group of birds or bats gathered to sleep, or the place where they are roosting.

Some birds roost in very large numbers.

In a large roost, snuggling up close together allows birds to:

- Share body heat and keep warm, which is extra important in winter.
- Have safety in numbers.
- Share information about good places to feed.

As well as starlings, look out for roosts of pied wagtails, often on trees or buildings in towns. Listen for them calling overhead, look out for lots of poo on the ground and look up to see the birds.

An incredible 20 million Mexican free-tailed bats roost in Bracken Cave in Mexico during the breeding season.

A perching bird sleeps standing on one leg, with its head tucked under its wing and the other leg snuggled into its belly. It can do this because of the way the tendons in its feet are arranged. When the bird's weight presses down on the foot, its perching grip is locked in position.

Did you know?

Swifts and frigatebirds sleep on the wing. We think they do this by sleeping with half of their brain at a time, like dolphins.

Wetlands can be roosts for thousands or tens of thousands of birds in winter. Wading birds such as knots, redshanks and sanderlings, as well as ducks, gulls and geese, can all sleep on the water or mud. Sometimes birds roost in a nest box. Families of long-tailed tits do this, while the world record is 61 wrens in a standard nest box!

Find out more: Starling murmuration, page 332

Tides

Animals that live on the coast deal with something else as well as day and night. They have to cope with the tides, which can create dramatic changes in their environment twice every day.

Tides are the rising and falling of the water level of the oceans. The Sun and Moon pull on the Earth as the Earth spins in space and this causes the oceans to bulge and recede. On the shore, the water moves inwards and upwards (rising, or flowing) and then outwards and downwards (falling, or ebbing). This means that part of the shoreline is underwater for some of the time and exposed for some of the time – a big change in conditions for the creatures that live there.

Tides rise and fall twice every day. What's more, they vary throughout the year. Each month has a spring tide, when

high tide is higher and low tide is lower than average, and a neap tide, when high tide isn't as high as usual and low tide isn't as low.

The shoreline is a great place to see wading birds. They find plenty of food there, such as molluscs, shrimps, crabs, worms, fish and seaweed. The animals and plants that live on the shore have adapted to survive in conditions that are salty and vary from completely submerged to dry and windswept.

Did you know?

Tides don't really move in and out on the shore. In fact, the bit of the Earth you are standing on moves in and out of the bulging tides!

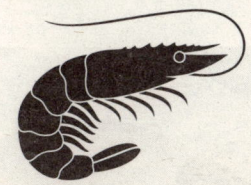

Rock pools

Rock pools are home to a variety of creatures. Mostly they are active when the pool is full of water at high tide, then they hide away and protect themselves from the dry, salty air and predators when the water drains away at low tide. You might find . . .

Limpets – hold very tight to the rock at low tide. At high tide they move around the rock, scraping up algae. They always return to exactly the same spot and are so precise with

this that they end up making a circular dent in the rock.

Whelks – spiral-shaped snails that eat other shelled creatures such as mussels and barnacles.

Barnacles – tiny, cone-shaped shelled animals that cling on to rocks, wood and even to other animals. They have tiny feathery legs that they wave at high tide to catch tiny specks of food.

Sea star (sometimes called starfish) – the top predator of a rock pool environment. At high tide it soaks up water into its body, which keeps it cool and damp during the hot and dry conditions of low tide.

Sea anemones – at high tide these soft-bodied animals wave their stinging tentacles to catch food. At low tide they squish up into a jellylike blob.

Shore crabs – have a hard shell like a pie crust that protects them from drying out or from getting smashed on the rocks by waves. They scuttle about looking for food when the tide is moving in and out.

Blennies – these
fish can actually walk
across the rocks
to find a good, wet
place to hide during
low tide.

Oystercatchers – explore the rocks and cracks
picking limpets, crabs and cockles to eat. Individuals
have different feeding styles – some stab a mollusc and
prise it apart, others
smash an animal's
shell, while
others use
their bills
like tweezers
to pick up
worms.

Wildlife through the year

January

Look and listen for a puffed-up robin, one of the few birds to sing all winter

Spot winter visitors such as redwings and fieldfares, especially if there is a spell of very cold weather

February

Look out for hazel catkins and for bees feeding on this early source of pollen

Birds are pairing up, doing their courtship displays and building nests

March

Look out for your first swallows and listen out for chiffchaffs and even the cuckoo

Enjoy different types of blossom, and look out for early butterflies

Keep an eye out for frogspawn, ducklings and displaying newts

Birds start to lay or hatch eggs

Red deer antlers have dropped and you can go out and collect them!

April

Bats come out of hibernation and you can see them feeding at night

Carpets of bluebells brighten up ancient woodlands

This is lekking month for capercaillie

The first white-tailed eagle egg hatches

May

Listen to the dawn chorus at its peak

Look out for frog and toad tadpoles in your nearest pond

Swifts are one of the last summer visitors to arrive, speeding and screaming overhead (they are quick to leave, too, and will be gone by the end of August)

Look for fawns sitting quietly

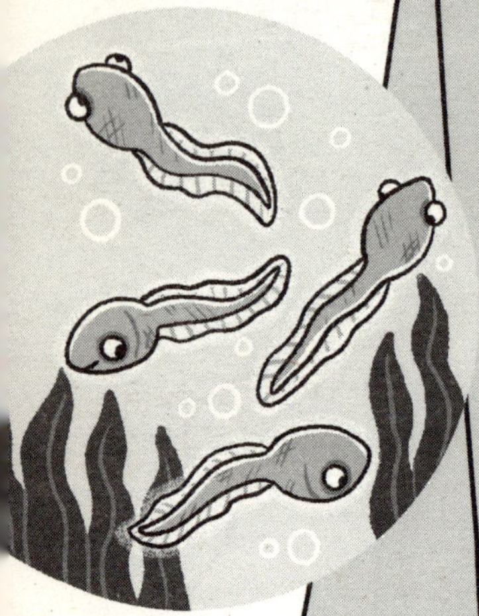

June

Fledglings are taking their first hops and flutters out of the nest

Seabird colonies are at their busiest

Look out for jewel-like dragonflies and damselflies

July

Long summer evenings are a good time to spot the hummingbird hawkmoth

Swarms of flying ants leave their nests over the course of several weeks

This is usually a bumper month for butterflies

August

The summer holiday is a good opportunity to explore rock pools (although they're fascinating at any time of year)

September

Woodlands and hedges will be full of berries, which are food for birds, squirrels, mice and pine martens

Jays are busy burying acorns

Wasps develop a taste for sugar and buzz round rotting fruit as well as any sugary human food such as jam or sugary drinks

October

Listen for tawny owls in the evening – they are noisiest this month

Give hedgehogs a warm, safe place to hibernate, such as a pile of leaves or even a hedgehog house

Bees are still busy – you can watch them and other insects feeding on ivy plants

This is the month of the red deer rut

November

Birds' natural food sources are getting thin, so this is a great time to attract them to your garden

This is one of the best months to visit a wetland nature reserve, where thousands of birds are gathered

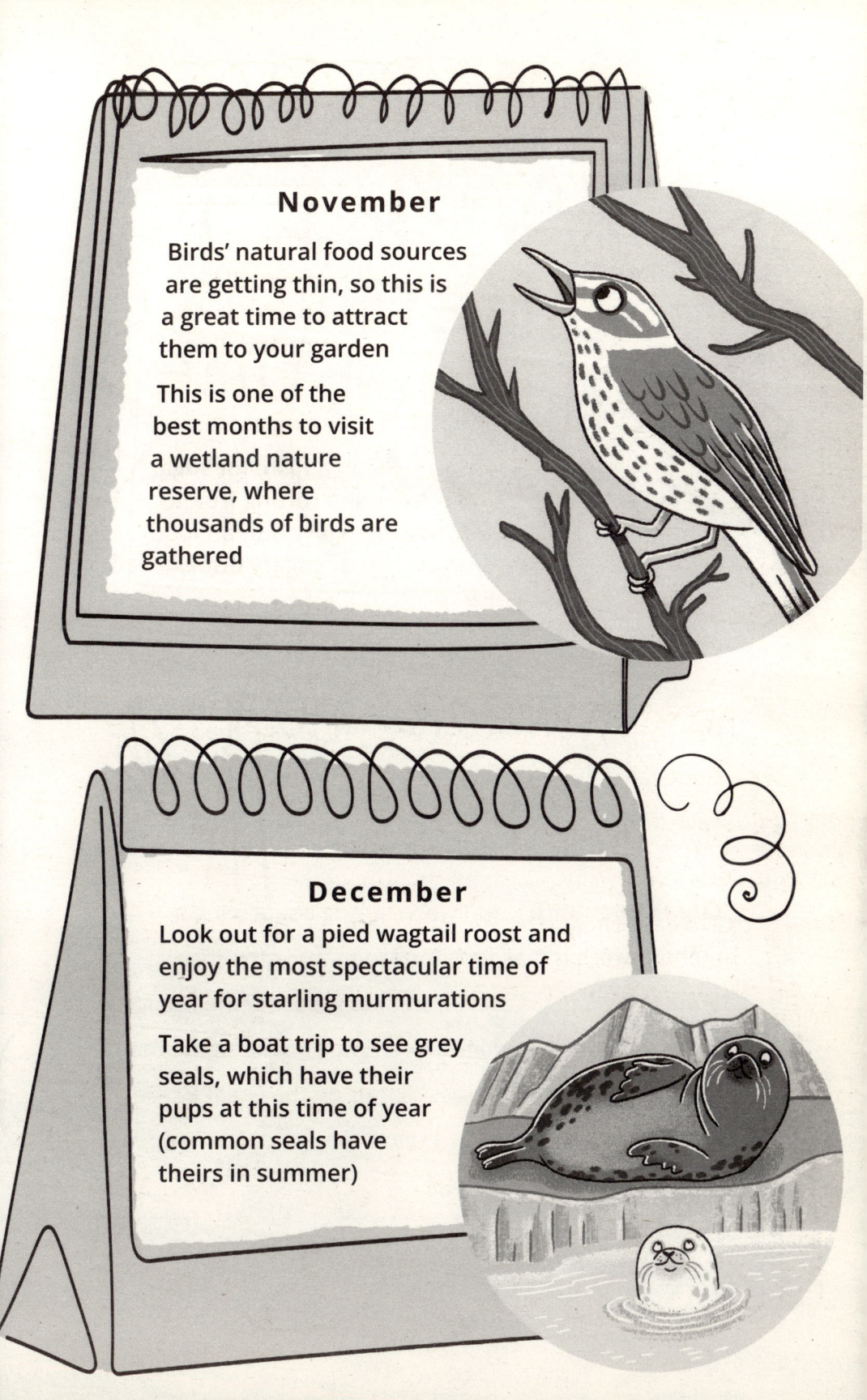

December

Look out for a pied wagtail roost and enjoy the most spectacular time of year for starling murmurations

Take a boat trip to see grey seals, which have their pups at this time of year (common seals have theirs in summer)

Different periods of time

Animals live according to the rhythms of different periods of time.

Day by day they wake and sleep in line with their body clock. Each day the Sun rises and sets. In coastal areas, each day brings two high tides and two low tides.

Month by month – and we're talking about a lunar month, which is how long the Moon takes to move around the Earth – coastal animals adjust to the changing levels of the high and low tides. Some animals time their breeding by monthly cycles. Corals on the Great Barrier Reef in Australia breed in a mass spawning that happens during one full moon a year.

Year by year many animals breed, many animals migrate, and some hibernate. (Not all animals breed in yearly cycles, though – some do so at any time of year or are triggered by events such as heavy rainfall.)

Animals adapt and change to succeed in different environments and conditions. This is called **evolution**. It takes many generations and generally happens over very long periods of time, sometimes millions of years.

Sometimes, evolution happens much more quickly. Animals that live in a rapidly changing environment have more pressure to adapt and change. Italian wall lizards introduced to a tiny island in Croatia just 30 years ago are evolving fast. Already they have bigger heads, a larger bite and a different gut structure from other Italian wall lizards, allowing them to eat plants rather than insects.

Life on Earth clock

Life on Earth has existed for so long that it's mind-bending to imagine it. In relation to all life, humans have been alive for the blink of an eye.

If you imagine the Earth's whole history (4.6 billion years) squeezed into a single day . . .

 around 4 a.m. a simple, single-celled life appears

 dinosaurs live from around 10.40 p.m. to 11.40 p.m.

 the first mammals appear at around 11.39 p.m.

 early humans appear at around 11.58 p.m.

 modern humans (like you and me) appear just before the stroke of midnight

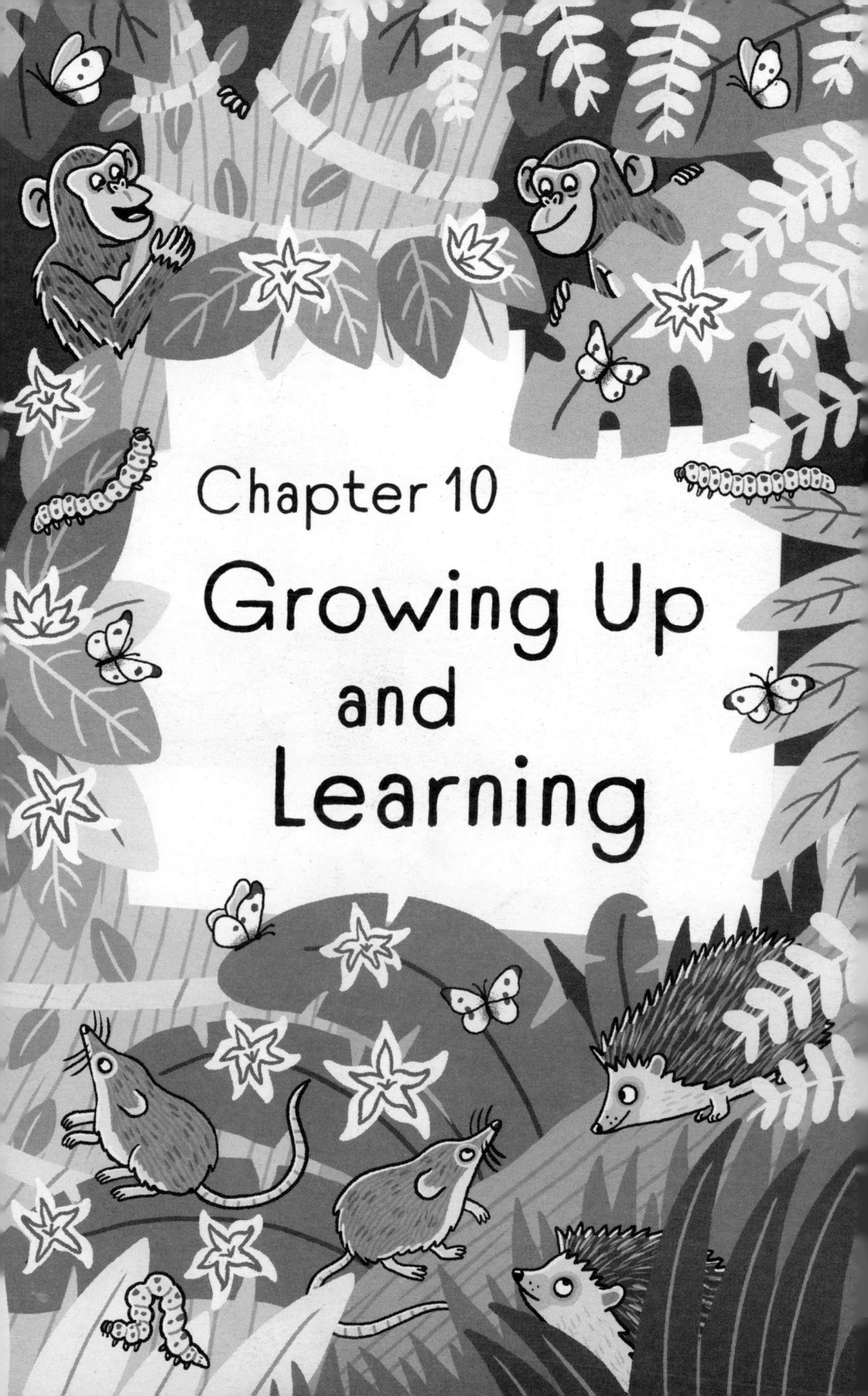

Chapter 10
Growing Up and Learning

1

And in at number one, my all-time favourite animal is . . . drum roll . . . the **mountain gorilla**.

Why the gorilla? Well, they're big, they're powerful, they're strong.

A mature adult male gorilla is called a silverback. He is huge – about twice the weight of an average adult man and a lot bigger than a female gorilla. It's pretty impressive to get that sort of weight on a diet of mainly plants, with a few ants and termites! The silverback is the boss of the troop and defends it from outsiders. Usually in a troop of gorillas there is just one silverback, but if there are more there is a clear number one, number two, number three and so on.

When a male becomes
a silverback, at about 12
years old, he develops a
silver-grey saddle across
his back and thighs. He has
huge, strong arm muscles and a
bulging bony crown on his forehead.

Gorillas fight completely differently from chimpanzees.
The chimps will get aggressive – they'll make a racket
and smack everything – over very little. A gorilla, on the
other hand, will just look at you. It's as if he's saying, 'Do
we have to go through this again? No, let's move on.' It
literally is just a look, and that's enough to tell another
gorilla not to mess with him. Sometimes he will push over
a small tree just to show his strength and power.

For a silverback, success is how many babies he has.
He needs to protect them above everything. He guards
his family each night. He sleeps on the ground at the
bottom of a tree, and sends his family up to sleep in the

365

branches. There, they make nests to sleep on. Sometimes they'll even cover themselves with branches, like bed covers, because it's going to rain. The silverback knows he's too heavy to go up the tree with them.

The young gorillas of the troop learn from their parents and family members. They climb trees, learn to forage, have play fights and explore the forest together. When they grow up, all male gorillas and most of the females leave the troop and find another troop to join.

If you're lucky enough to go and see gorillas in the wild, you mustn't look them in the eye. If one looks at you, you look to the floor. When they're comfortable, they make a noise like like us clearing our throats.

Gorillas are such incredible animals. It's very difficult to see them in the wild because they live in remote tropical forests, and today many of the places they live in are unfortunately in the midst of wars. It's not good for the gorillas to come into contact with humans too much. They can get too used to us and may not avoid poachers. They can also catch our diseases, from common colds to coronavirus. And they are endangered – threatened by habitat loss, disease and poaching.

But conservationists are working hard to protect them and help their populations to recover. There has been some good news in recent years, with populations increasing slightly. I really hope I get to film them one day.

367

Facts

Scientific name:	*Gorilla beringei beringei*
Mammal family:	great apes
Height:	120–180 cm to shoulder
Weight:	up to 200 kg
Found in:	tropical forests in Africa
Eats:	mainly juicy plants – stems, berries, leaves and bark, sometimes ants and termites
Babies:	between 4 and 6 over a lifetime
My three words:	family, intelligent, powerful

Gorillas spend about a quarter of the day eating. Food is plentiful in the tropical forests where they live, so it is easy to find and they don't have to fight over it.

Gorillas are highly intelligent. Like us, they have worked out how to use tools. Sometimes they use sticks to work out how deep a stream is, use twigs like cutlery to scoop up insects to eat or make ladders out of bamboo to help babies climb up to their treetop nests. One gorilla in Congo was seen using a small branch as a walking stick to help steady her as she waded through a pool.

Gorillas use their hands almost exactly as humans do. Many animals use their mouths to move objects, but gorillas use their hands.

Gorillas share 98.3 per cent of their genetic code with humans. Chimpanzees (98.8 per cent) and bonobos (98.7 per cent) are our even closer cousins.

Gorillas have little teeth behind their incisors, specially for stripping the bark off little twigs.

Growing up

Some animals grow up fast and are old enough to have their own babies when they're just six weeks old. At the other extreme is the female Greenland shark, which doesn't become an adult until it is at least 134 years old.

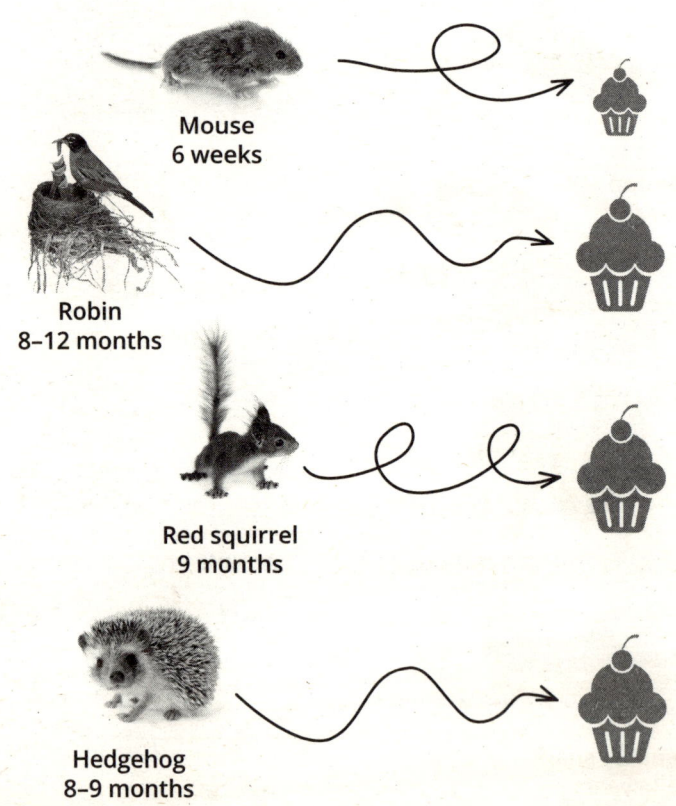

Mouse
6 weeks

Robin
8–12 months

Red squirrel
9 months

Hedgehog
8–9 months

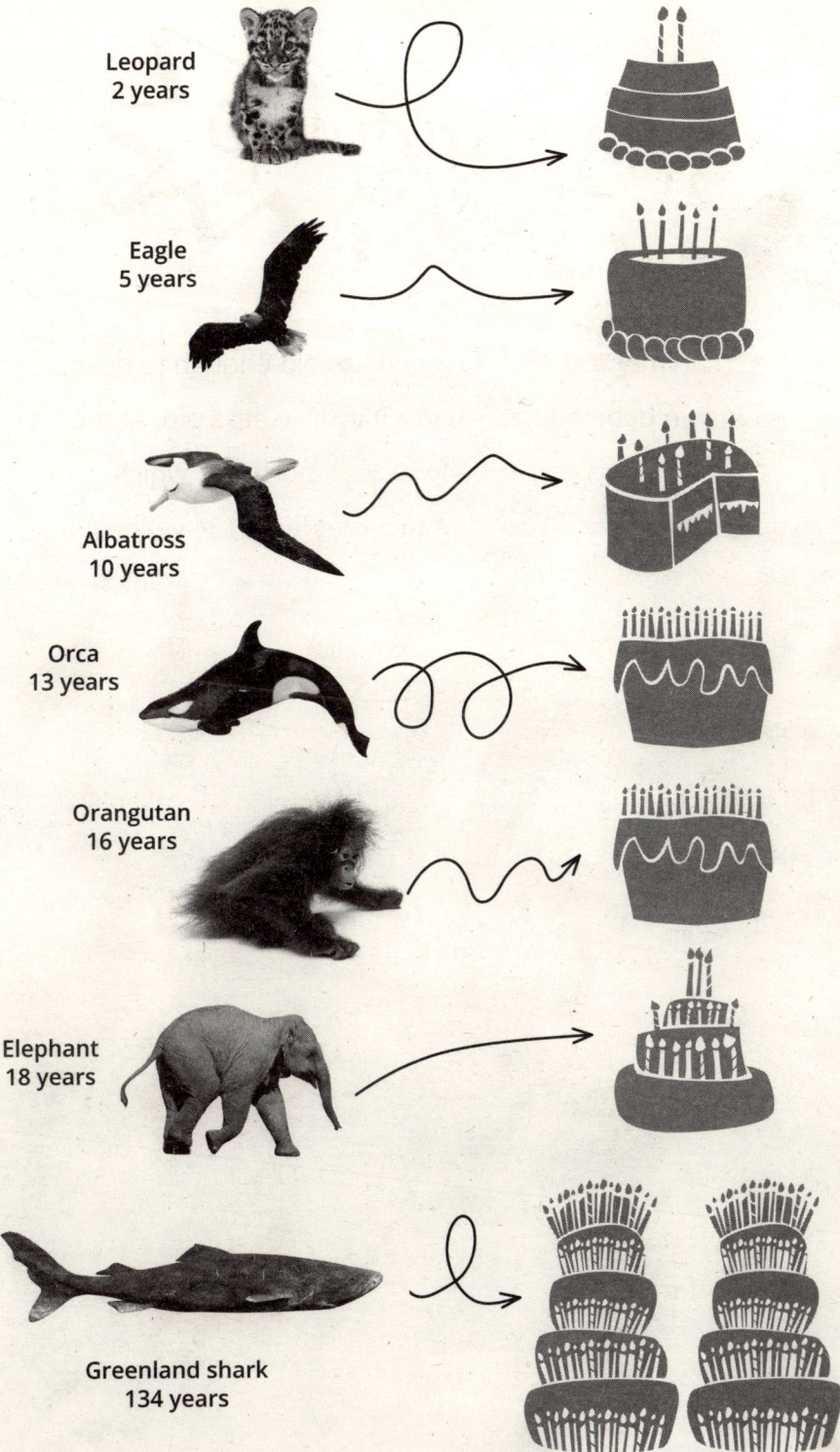

Leopard
2 years

Eagle
5 years

Albatross
10 years

Orca
13 years

Orangutan
16 years

Elephant
18 years

Greenland shark
134 years

Orangutan – forest home

Being an orangutan mother is a full-time job for as long as seven years. An **orangutan mum** . . .

- ◎ Has a detailed **mental map** of her rainforest home, so she knows where the best food is to be found at all the different times of year. It takes several years for a baby to learn this knowledge from its mother.

- ◎ Carries her infant **on her back.**

- ◎ Can eat using her **feet** as hands – useful if one hand is holding on to a branch and the other is cradling an infant!

- ◎ Makes a **new nest** for herself and her infant every night. She makes a sleeping platform by weaving branches together and adds more branches for a mattress and – if it's rainy – a roof.

- ◎ **Nurses** her infant (feeds it on milk) for five years.

- ◎ **Stays** with her baby, even when the youngster is walking and climbing on its own, for seven years – the longest childhood of any animal except humans.

◎ Only has a **new baby** once every seven to nine years – the longest gap of any land mammal.

◎ Lets her daughters stay with her when a new baby arrives so they can learn **parenting** skills from her.

Orangutans are under serious threat because of their rainforests being destroyed, often for palm oil plantations. If mothers are killed, the orphan babies can be cared for in wildlife sanctuaries but sadly they don't have a mother to teach them how to survive. Orphan sanctuaries provide 'forest schools' where foster mothers take care of them and lead them around the forest.

Orangutan is the Malay word for 'person of the forest'.

Nature jobs ...

There are people working with nature and for nature even in the middle of a city.

I'm a city planner. I make sure that homes, offices, shops, streets and transport are built in a way that allows for nature.

I'm an architect and I make sure that trees are part of my designs, so that there are places where wildlife can live.

I'm a builder and I use sustainable materials and build systems that save water and energy.

I'm a wildlife officer. I help others make space for wildlife in parks and gardens – for example, putting up bird boxes and bat boxes.

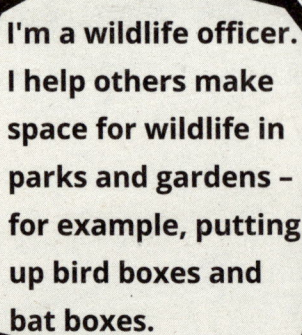

I'm encouraging my neighbours to build hedgehog highways.

I'm a politician. I help create laws that help nature and make sure that cities, towns, farmers and ordinary citizens have money and support to protect nature.

I've set up a conservation club at my school. We're creating a wild space for nature and spotting wildlife in our city.

In the countryside, farmers work with nature alongside growing our food. They look after large areas of land and can manage it in a way that is good for nature. For example, they can keep the use of pesticides and other poisonous chemicals to a minimum. They can help improve biodiversity by allowing space for hedges and other wild areas, and they can make sure the water and soil in an area is good quality and not polluted.

Metamorphosis

Most animals grow up in quite a straightforward way. They start life as babies that look like small versions of their parents then grow and develop until eventually they are adults.

Some animals, however, have a complete change of shape as they grow up. A caterpillar, for example, doesn't look anything like a butterfly. Its body shape is completely different. This kind of change is called metamorphosis.

Butterfly

A butterfly starts life as an egg.

This hatches into a larva, called a caterpillar.

The caterpillar grows and grows. Eventually it forms a chrysalis around itself – this is like a cosy sleeping bag for the caterpillar.

At last the butterfly emerges from the chrysalis.

In the hard shell of the chrysalis its body completely disintegrates, turning into a soupy mush, and reorganizes itself into the shape of an adult butterfly.

Frog

Frogs start life as an egg, known as frogspawn.

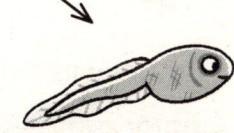

Each egg hatches into a tadpole. At first the tadpole has a tail, no legs, and breathes through gills in the water. The tadpole changes shape a few times.

Then it absorbs its tail into its body and becomes a froglet.

Finally it hops out of the water.

First it loses its gills and grows back legs.

Next it grows front legs and its head becomes more frog-shaped.

Toads, newts and salamanders follow the same stages. Newt dipping is one of the best things ever. Go looking for them in a pond at night – ask an adult to go with you. Shine a torch into the water and you can see perfectly into their world as they interact with each other and shimmer their tails to attract a mate.

Animal siblings

Do you ever hold hands with your brother or sister? **Common shrews** sometimes hold tails! They follow their mother in a line forming a 'caravan', with one holding another's tail.

Young **peregrine falcon** siblings play together, practising their hunting skills, with one pretending to be the hunter and the other pretending to be the prey.

In **Asian short-clawed otter** families, the older siblings help raise the young pups.

All the **worker bees** in a hive are sisters. They work together, feed together and look after the eggs and larvae – which are also their younger siblings – together.

Lions, chimpanzees, elephants and orcas have close sibling relationships. A study in the 1980s found that some orca brothers never spend more than a few hours apart over their whole life. In fact, they help raise their

brothers, sisters, nephews and nieces, but never take care of or even meet their own offspring.

You might have seen flocks of tiny **long-tailed tits** flitting around a garden, park or woodland. The flocks are usually family groups. At breeding time, adult helpers join in with feeding and looking after the new chicks. These helpers are most often brothers of the parent bird. So uncles play a vital role in the family. Sometimes there are as many as ten chicks, so the parents need all the help they can get!

Some animals are terrible siblings. It's quite common for baby animals to kill their brothers and sisters to make sure they get all the food and care themselves. **Golden eagles**, **hyenas**, **cattle egrets** and **blue-footed boobies** all do this. **Sand tiger sharks** even eat their own brothers and sisters while they're still in their mother's body!

How do animals know what to do?

How does a polar bear know how to hunt? How does a bird know how to fly? How does a beaver know how to build a lodge? How do any animals know what to do? It comes down to a combination of things.

Instinct

Instinctive behaviour is behaviour that is automatic and inherited – in the animal's genes. They are born already knowing how to do something.

◎ **Cygnets know to follow their parents (this instinct is common to many baby animals and is called imprinting).**

- Robins instinctively attack things with a patch of red – it could be a rival robin or it could be anything!
- Herring gulls automatically sick up food to feed their chicks when one pecks the red spot on their beak.

Stimulus

A stimulus is a trigger from the outside world that tells an animal to do something.

- When days start getting longer in spring and birds sense the change in daylight, this is the trigger for them to get ready to breed.
- When polar bears hear the sound of ice cracking, they flee from the sound.
- When the weather gets cold and food supplies run out, hedgehogs know it is time to hibernate.

Learning

Like us, all animals learn. They watch their parents or other animals, they learn from observing and experiencing the world, and from trial and error.

Find out more: Learning, page 386

Playtime

If you've ever had a kitten or puppy at home, you'll know how much they love to play. Play is important for many young mammals as it's one of the ways they learn the skills they need to survive in life.

A baby elephant plays with its trunk, whirling it and wiggling it, because it hasn't learned all the amazing things it can do with it yet.

A lot of animal playtime consists of play-fighting or pretend hunting. You see the young animal pushing their weight around. They're just trying to figure out, how strong am I?

In the young male polar bears we see it. They don't want to injure themselves, but they want to find out their strength. We call it sparring. It's practising, like when a human boxer spars. Polar bears love to spar in the water, because they're a bit more floaty, it's nice and cool and they're well fed. They're not going to spar on a hot day. They push each other and fight each other in the water. Then later in the year, when the snow comes but the sea ice hasn't fully frozen up yet – that's when they start sparring on land. That's starting to build up to a real fight, because by this point they're ready to mate, their mums have given birth to new cubs, they're on their own in the world.

You see young animals testing their weight –
whether that's a polar bear or a lion or a gorilla
or a chimpanzee, you see them pushing on
something: can I break this piece of ice, this
log, this stone? When they get old enough,
this will be something they do for real,
when they're finding food.

When I was studying chimps in the Welsh Mountain Zoo it was interesting to see some of the young ones pushing their weight. There was one juvenile, Euro, who would start a fight just so she could learn who to mess with and who not to mess with. When everyone else was asleep, she'd go to the biggest male she could find, pick up her hands and just slap them down on his back. The male would wake up – angry, ready for a fight, but see the baby and obviously couldn't beat her up so would let off his aggression on another adult. Winding him up to see what happens – that's a lesson for her to learn.

Cats and kittens stalk prey, and if something runs or makes a quick movement, their instinct is to chase it. So a kitten chasing a toy or a string is acting on instinct but developing its hunting skills.

Most animals that play are doing so as a way to learn life skills. But dolphins also love to play just for the sake of it. Even the adults play, whether it's riding the bow wave in front of a boat, picking up a bit of seaweed and dancing with it, or creating rings of bubbles in the water and bursting them by biting or swimming through them. Perhaps, like us when we play, they are exercising their intelligence and being sociable with their friends.

Learning

Animals have a lot of life skills to learn. They learn to hunt, to fly, to build, to make a nest, to impress, to make tools and even how to breed.

 When fox cubs are learning to hunt, they practise how hard to bite. You see them play-fighting with their brothers and sisters, biting each other's tails, pouncing and nipping, until the mother comes and smacks them. She's also teaching them to hunt by bringing them food torn into little bits. The cubs

play with the bits of food – chucking a piece up in the air and pouncing on it, so they're learning to take down moving prey.

◎ Birds don't really play-fight but some birds of prey practise hunting with their siblings. They fly and roll together, one tries to catch the other, hold each other's talons, and let go. When they hunt for real, they will clench their talons to grip their prey so they're practising this action.

◎ When eagles are in the nest, you see them practising flapping their wings. They're building up their flight muscles to get ready for flight.

◎ Young red deer stags will join in the rut and try to impress the females, but they also need to watch the more experienced males to learn how to put on an impressive display.

Birds learn from experience throughout their life. There are many things that might cause a nest to fail and the eggs or chicks to die. The nest may be too easy for predators to reach or vulnerable to being washed away or baked in the sun, there might not be enough food to feed the chicks, the parent bird may make mistakes such as leaving the eggs for too long so they get cold. First-time parents often don't succeed but with each nesting attempt the birds are learning to give them a better chance in future.

Chimpanzees learn to make tools by watching others do the same – they even deliberately share their knowledge by demonstrating to others what to do or giving their tools to others.

Hamza's Nature Heroes
Sir David Attenborough
Born 1926

Of all the naturalists and wildlife presenters I admire, Sir David Attenborough is the greatest. The boss. His *Life of Birds* was one of the first programmes I saw when I moved to England. His incredible knowledge and enthusiasm has opened up the natural world for millions of people, not just me.

Sir David grew up in Leicester, England, and spent his childhood discovering animals and hunting for fossils. After studying science at university, he started working at the BBC and became a TV producer. One of the first programmes he created was *Zoo Quest*, which showed animals in zoos and in the wild, in places viewers had not seen before. There had been some natural history

programmes before this, but many people would say this was when the idea really took off.

In the 1970s he moved on to writing and producing his own television series. *Life on Earth* was the first and he has created many incredible natural history series that have won numerous awards and shown the wonders of nature to millions of people.

One very famous sequence showed David meeting mountain gorillas in Rwanda in the 1980s. The scene starts off with him doing a piece to camera and then one of the babies comes up and goes like, 'Hey, what's up, my man? How you doing?' Then everyone's looking around at

the silverbacks, concerned, but they all decide, 'Yeah, he's cool, he's fine.' Attenborough just speaks to the camera, and that's when all of the gorillas come in – because the baby is comfortable, the others are happy, too. You can see the babies playing with Attenborough, tickling him, climbing all over him, taking off his shoes, and all the time he's trying to deliver the piece to camera, although in the end the gorillas' behaviour says more than his words can. It was a very rare and special moment.

My own love for gorillas started when watching this famous Attenborough sequence. I thought, I want to meet them like that, have them climb over me and take off my shoes!

Sir David has always been involved in conservation and protecting wildlife. In recent years he has focused more strongly on getting across the message that nature is under threat from human activity and that we need to change things before it is too late.

Sir David has had dozens of species named after him, including a tiny spider, an echidna, a tropical butterfly, a carnivorous plant (see picture below), a blue Australian lizard and several prehistoric creatures, including an enormous prehistoric sea reptile, Attenborousaurus!

This is a fan-throated lizard from Kerala, India, called 'Sitana attenboroughii'.

Find out more: Gorillas, page 364

Memory

When you are learning, you need to practise your skills, but you also need to learn information and remember it. This is the same for animals.

Animals that migrate know when and where to go by instinct, helped by triggers such as day-length or temperature. But they hone their skills with each migration they make, building on experience and memory. They remember landmarks along the route and recall which areas were good for finding food or which were difficult to cross.

Tweety birds learn to sing by imitating others of their species and memorizing a template for their song. Over time, individual songbirds refine their song through practice. Some species have intricate songs and individuals make their own versions that are slightly different from others. They also build a 'songbook' of a few different tunes they can perform, depending on the occasion!

Grandmothers

In elephant herds, matriarchs are the leaders. The matriarch is an older female who is the grandmother or great-grandmother of all the other females and who holds a lot of important remembered knowledge. The vast plains where elephants live can sometimes become very dry, with food becoming scarce. Elephants can smell water and greenery from a long way away, but sometimes even that is not good enough. The grandmother remembers places that the herd may not have visited for many years. She will lead the herd to the right place and this allows them to survive drought.

In orcas, females can live to 80 years old. They spend their energy teaching and caring for their children and grandchildren, and passing on their experience and knowledge long after they've finished having calves themselves.

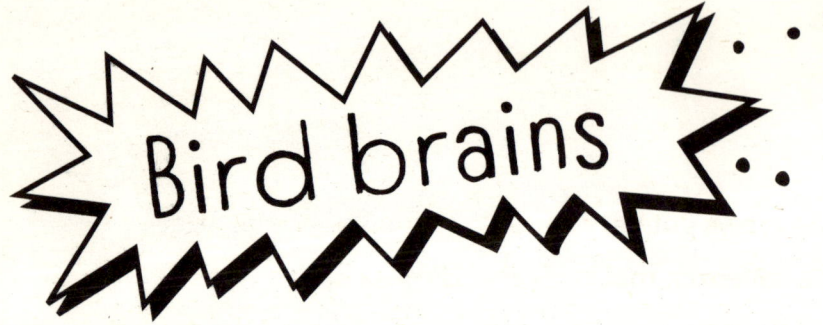

Bird brains

The prize for the top learners in the bird world goes to the corvids (the crow family).

◎ **Crows can remember human faces. In an experiment at the University of Washington, scientists wore caveman masks to trap wild crows for ringing. Several years later, the scientists put on the masks again, along with others wearing masks of famous people, and set out to see the crows.**

I've seen you before . . .

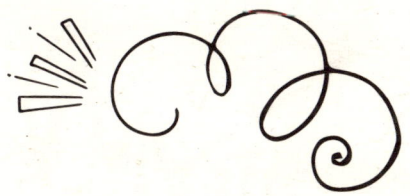

The crows got very agitated when they recognized the caveman masks. Some of these crows were not the same as the original birds, which suggests the original crows warned the others about the 'dangerous' faces.

◎ **Corvids** can use tools and solve problems. They often use sticks and stones to get at food. Like **chimpanzees**, **New Caledonian crows** often take a stick, strip off the leaves and bend it to make a hook. In experiments, **ravens** have solved multi-stage puzzles and challenges using different tools to eventually get at food.

◎ Corvids plan for the future. In experiments, **ravens** have stored tools to use later. **Jays** 'plant' acorns, ready to feed the saplings to their young.

◎ **Ravens** show empathy. Scientists have observed them consoling other ravens who have been in a fight.

Find out more: Chimpanzees, page 176; jays and acorns, page 197

397

How to get involved

So, you've read this book, and now you want to get more involved in nature.

As Erin and Gemma said, the most important thing to do is to get outside! Get to know the wildlife on your doorstep – in your local area and nearby parks, woodlands, rivers and wild spaces.

When you're out and about, really notice what is going on. Look and listen for birds, get down close to the ground and examine the flowers, grass and insects and look for clues that animals have left behind. Be still and wait for the wildlife to show itself.

Take photos or videos, draw what you see or write it down. If photography is your thing, just take as many photos as you possibly can. Get to know how your camera (or phone) works and practise taking different kinds of pictures. Try

398

close-ups of flowers, butterflies, bees or bugs. Practise shooting ducks, swans and pigeons in your local park – they're easier to photograph than little tweety birds.

Sometimes you'll want to visit a nature reserve. These can be fantastic places to see creatures you haven't seen before. Talk to other birdwatchers, who can give you tips on how to spot and recognize different birds and may even let you have a look through their telescope. This is how I learned a lot about birds. My Mum would take me to see the ospreys at Rutland Water, and I would talk to the experienced birdwatchers there.

Always stay safe when you're out in nature. Make sure an adult knows where you are going and when you'll be back, and take care around roads and water. Or ask an adult to go with you. Make sure you dress for the weather, too.

Always be respectful of nature and take care not to damage habitats or creatures. Don't make too much noise or leave litter behind.

And on the days you can't get out into nature, you can immerse yourself in wildlife programmes on TV. Why not start like I did, with the wonderful work of Sir David Attenborough.

Here are some clubs and organizations you can join:

Bumblebee Conservation Trust
bumblebeeconservation.org

RSPB
rspb.org.uk

Butterfly Conservation
butterfly-conservation.org

The Wildlife Trusts
wildlifetrusts.org

Buglife
buglife.org.uk

World Wildlife Fund
worldwildlife.org

Jane Goodall Institute
janegoodall.org

Wildfowl and Wetland Trust
wwt.org.uk

Useful sources of information:

*Merlin bird identification app, available from the App store

*How to examine an owl pellet:
discoverwildlife.com/how-to/identify-wildlife/owl-pellets-identify-dissect
youtube.com/watch?v=aY1zsBH0vnk

*How to identify animal tracks:
wildlifetrusts.org/how-identify/identify-tracks

*If you find an injured animal:
Royal Society for the Prevention of Cruelty to Animals
www.rspca.org.uk
Scottish Society for the Prevention of Cruelty to Animals
www.scottishspca.org
British Divers Marine Life Rescue (BDMLR) Rescue Hotline
bdmlr.org.uk

400

Glossary

aerodynamic having a shape that allows a thing to move through the air smoothly and easily.

amphibian one of a group of animals that includes toads, frogs, newts and salamanders.

apex predator an animal that is the top predator in its habitat, with no natural predators of its own.

binocular vision seeing with both eyes facing front, which creates an image that allows an animal's brain to work out distance.

biodiversity the variety of different living things in a particular place or of the whole Earth.

bioluminescent able to create glow-in-the-dark light.

cartilage a strong, flexible substance in the body. Sharks, skates and rays have skeletons made of cartilage. We have cartilage in between the bones in our skeleton – it acts like a cushion.

clutch the name for a nest of eggs or baby birds.

compound eyes eyes made up of thousands of individual lenses. Animals with compound eyes, such as many insects, see the world a bit like a mosaic, can spot objects super quickly and have almost total all-round vision.

cold-blooded needing warmth from the Sun to warm up each day. Reptiles, amphibians and most fish are cold-blooded.

decomposers living things that break down dead material so that its nutrients can be recycled and provide goodness for plants and animals. Examples include bacteria, fungi and some worms.

detritivores animals that feed on dead material so that its nutrients can be recycled. Examples include woodlice, millipedes, slugs and snails, worms, beetles and mites.

dominant the most powerful or lead animal in a group, the boss; or, the most abundant and successful living thing in a particular habitat.

down very fine, soft feathers underneath a bird's strong outer feathers. Baby birds only have down feathers, before their other feathers have fully grown.

echolocation a method of using hearing to sense the location of objects by listening to echoes. Bats, dolphins and some other animals use echolocation.

ecosystem engineer a living thing that has such an important effect on the environment around them and the other wildlife in it that it can be said to shape and create that environment. Ecosystem is a word for a network of living things that are all linked together and linked to their environment.

electrosensitive able to sense the faint electric pulses given off by other living things.

energy efficient using relatively little energy to produce plenty of power.

evolution the process by which living things adapt and change over time.

habitat the place a living thing inhabits, where it can find food, shelter and a safe place to breed.

holt an otter's nursery den, where it has its babies.

imprinting a baby animal's instinct to follow and stick close to its parents.

incubate sit on eggs to keep them warm until they hatch.

infrared a kind of light energy. Human eyes cannot see infrared but some animals, including some snakes, can.

infrasonic sounds that are lower than humans can hear.

instinct behaviour that animals are born already knowing how to do.

invertebrate one of a large group of animals with no backbone.

iridescent a shiny, shimmery effect such as on a starling's or butterfly's wing. Iridescence is not a colour but a way that light bounces off something and creates the impression of changing colour.

keystone species a living thing that is so important to an ecosystem that without it the whole ecosystem starts to collapse and biodiversity is lost.

kopje a small rocky hill rising above the mainly flat African grassland.

litter a family of young animals born to one mum at the same time.

magnetic sense ability to sense the Earth's magnetic field.

mammal one of a group of animals that are warm-blooded, have fur or hair, four limbs and feed their young on milk made in the mother's body.

mate a mate is the partner an animal breeds with. To mate is to come together to breed.

matriarch an older female in a group of animals, who leads the group.

matriline a group of animals that are all related to one female, for example she is the mother or grandmother of the others in the group. Orca groups are called matrilines.

metamorphosis a complete change of body shape as an animal grows up.

monocular vision seeing with eyes that work independently of each other. An animal with monocular vision gets a wide view around them, with each eye seeing part of that view.

nutrients substances that living things need to help them survive and grow. Animals get nutrients from food, whereas plants get nutrients from the soil, and make them using sunlight.

olfactory cells sense cells that detect smelly chemicals.

passerine one of a large group of birds also known as songbirds or perching birds (or, as I call them, tweety birds).

plumage all of a bird's feathers.

poaching illegal hunting.

proprioception ability to sense where your body is in space.

reptile one of a group of animals that are cold-blooded and have scaly skin, including snakes, crocodiles, turtles and lizards.

roost the word for sleep for birds and bats. A roost is also a group of birds or bats gathered to sleep, or the place where they are sleeping.

spawn the eggs of fish, amphibians or some invertebrates.

spraint poo, especially otter poo.

territory an area that an animal defends as its own, making sure that others keep out. It can defend its territory using songs and calls or by leaving scent or other markings.

ultraviolet a kind of light energy. Human eyes cannot see ultraviolet but some animals, including many insects and birds, can.

venomous able to inject venom (poison) into prey, for example by biting or stinging.

vertebrate one of a large group of animals with a backbone. Fish, amphibians, reptiles, birds and mammals are vertebrates.

warm-blooded able to make warmth inside the body. Birds, mammals and a few large fish such as some sharks and tuna are warm-blooded.

With thanks to

Jesse Wilkinson, Dr Gemma Clucas and Erin Ranney – thank you so much for letting me interview you for the book, you are my nature heroes.

Catherine Brereton for brilliant animal fact swapping and book creation.

Louise Forshaw for her amazing artwork.

And huge thanks to the brilliant team at Macmillan Children's Books, particularly my editor Gaby Morgan and Clare Hall-Craggs, Bethan Thomas, Charlie Morris, Sarah Clarke, Amy Boxshall, Tanny Hossain, Rachel Vale, Tracey Ridgewell, Sue Mason, and Farzana Adlington.

Everyone at DML Talent – my agent Louise Leftwich, Jan Croxson, Borra Garson and Megan Page.

Lastly thank you to my family and community of Kilchoan.